Iron Making In The Olden Times:

as instanced in the
Ancient Mines, Forges, And Furnaces
Of The Forest Of Dean,
Historically Related, On The Basis Of
Contemporary Records And Exact
Local Investigation:

By

Rev. H. G. NICHOLLS, M.A.,

AUTHOR OF "AN HISTORICAL AND DESCRIPTIVE ACCOUNT OF
THE FOREST OF DEAN," AND
"THE PERSONALITIES OF THE FOREST OF DEAN."

1866

ISBN: 978-1-78139-007-8

© 2011 Benediction Classics, Oxford.

Contents

DEDICATORY PREFACE. i

THE OLD "BLACK COUNTRY" OF GLOUCESTERSHIRE; OR, AN HISTORICAL RELATION OF THE MINING AND MAKING OF IRON IN THE FOREST OF DEAN, FROM THE EARLIEST TIMES. 1

A SURVEY OF THE FOREST OF DEAN IRON WORKS IN 1635. 27

AN ACCOUNT OF IRON ORE RAISED IN DEAN FOREST AND HUNDRED OF ST. BRIAVEL'S FROM CHRISTMAS, 1863, TO CHRISTMAS, 1865. 54

THE MINERS LAWES AND PRIVILLEDGES. 60

DEDICATORY PREFACE.

The remarkable revival and development that has recently taken place in the Iron Works of the Forest of Dean, and the consequent improvement which has accrued to the district, proves conclusively that its condition and prospects are largely dependent upon such manufacture. Impressed with this fact, it has occurred to the Author that a more particular account of them than has been given in his former work on the Forest might prove interesting to the numerous individuals with whom they are connected.

For several years past this subject has been upon his mind, during which time he has fully availed himself of the contents of the Forestal Archives belonging to the Middle Ages, and appropriated all the information, as he believes, which the neighbourhood itself affords.

He respectfully submits the produce to the perusal of those gentlemen and friends who may favour these pages with their attention.

In coming before them for the third time, he cannot retire from so interesting a neighbourhood without requesting them to consider this as his final mark of appreciation and gratitude for the invariable kindness they have so long shown him.

H. G. N.

April, 1866.

THE OLD "BLACK COUNTRY"
OF GLOUCESTERSHIRE;
OR,
AN HISTORICAL RELATION OF THE MINING
AND
MAKING OF IRON IN THE FOREST OF DEAN,
FROM THE EARLIEST TIMES.

If there be one circumstance more than another that has conferred celebrity on the Forest of Dean, it is *the remote origin, perpetuation, and invariably high repute of its iron works.* Uniting these characteristics in one, it probably surpasses every other spot in Great Britain.

In the author's former "historical account" of this neighbourhood, he gave all the information he had then collected relative to the mining and making of iron therein. Since that time, he has greatly extended his investigations, especially [1] amongst the records of the Court of Exchequer. The result is, that he believes he is now enabled to present to the public the most complete description that has yet appeared of the manufacture of iron during the Middle Ages, detailing, in the first place, all the particulars he has gathered of the operations of the primitive miner, or iron worker, and proceeding, in chronological order, to the present time.

In the year 1780, wrote Mr. Wyrrall, in his valuable MS. on the ancient iron works of the Forest:—

---

[1] Aided by the skilful labours of Stuart A. Moore, Esq.

Iron Making in the Olden Times

"There are, deep in the earth, vast caverns scooped out by men's hands, and large as the aisles of churches; and on its surface are extensive labyrinths worked among the rocks, and now long since overgrown with woods, which whosoever traces them must see with astonishment, and incline to think them to have been the work of armies rather than of private labourers. They certainly were the toil of many centuries, and this perhaps before they thought of searching in the bowels of the earth for their ore—whither, however, they at length naturally pursued the veins, as they found them to be exhausted near the surface."

Ancient Mine Work near Bream, commonly called "The Devil's Chapel."

Such were the remains, as they existed in his day, of the original iron mines of this locality; and, except where modern operations have obliterated them, such they continue to the present time.

The fact of their presenting no trace of engineering skill, or of the use of any kind of machinery, is conclusive of their remote antiquity. Nor are there any traces of gunpowder having been employed in them; but this, Mr. John Taylor says, was not resorted to for such purposes earlier than 1620, when some German miners, brought over by Prince Rupert, used it at Ecton, in Staffordshire.

It is the unanimous opinion of the neighbourhood that these caves owe their origin to the predecessors of that peculiar order of operatives known as "the free miners of the Forest of Dean;" a view which the authentic history of the district confirms.

They have the appearance either of spacious caves, as above Lydney and on the Doward Hill, or of deep stone quarries, as at the Scowles, near Bream. Or they consist of precipitous and irregularly shaped passages, left by the removal of the ore or mineral earth wherever it was found, and which was followed down, in some instances, for many hundreds of yards.

Openings were made to the surface according as the course of the mine-ore permitted, being softer to work than the limestone rock that contained it, thus securing efficient ventilation. Hence, although they have been so long deserted, the air in them is perfectly good. They are also quite dry—owing, probably, to their being drained by the new workings adjacent to them, and descending to a far greater depth.

In the first place, they were excavated as far down, no doubt, as the water permitted; that is, to a vertical depth of about 100 yards, or, in dry seasons, even lower, as may be seen by the watermarks left in some of them. Of these deeper workings, one of the most extensive occurs on the Lining Wood Hill, above Mitcheldean, and is well worth exploring. They are met with, however, on most sides of the Forest—in fact, wherever the ore crops out, giving the name of "meand," or mine, to such places.

Iron Making in the Olden Times

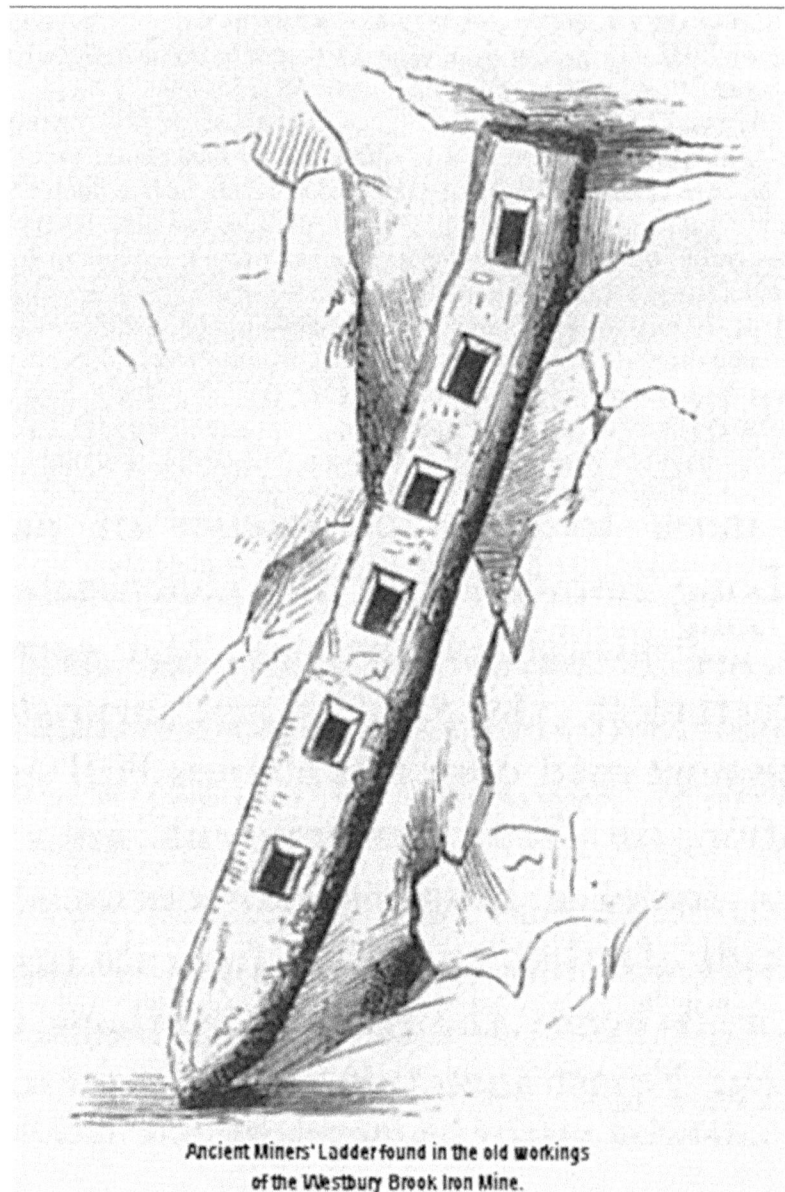

Ancient Miners' Ladder found in the old workings of the Westbury Brook Iron Mine.

Generally speaking, those spots where the ore lay exposed to view, would be apt to secure the notice of the earlier miners, and become the site of their more ancient workings. Not until they were pretty nearly exhausted would the severer labour involved in the lower diggings be resorted to. The shallower but more capacious mine holes appear with greater frequency on the south and west sides of the Forest, where, too, they were nearer to the water carriage of the Severn and the Wye. In most instances they are locally termed "scowles," a corruption, perhaps, of the British word "crowll," meaning cave. Occasionally they are found adorned with beautiful incrustations of the purest white, formed by springs of carbonate of lime, originating in the rocky walls of limestone around. Sometimes, after proceeding for a considerable distance closely confined in height and width, they suddenly open out into spacious vaults, fifteen feet each way, the site, probably, of some valuable "pocket" or "churn" of ore; and then again, where the supply was less abundant, narrowing into a width hardly sufficient to admit the human body. Now and then, the passage divides and unites again, or abruptly stops, turning off at a sharp angle, or, changing its level, shows rude steps cut in the rock, by which the old miners ascended or descended.

In some of these places, ladders, made out of hewn oak planks, with holes chopped through them for the feet, have been discovered. Mattocks, such as masons use, have likewise been met with, as well as oak shovels for collecting the ore. Shoe prints, and even shoe-leathers have also been found, although the latter are supposed to have been seldom used, judging from the more frequent occurrence of naked foot marks. Long immersion in the chalybeate water of the mine has blackened the oak, and corroded the iron; nevertheless, these relics are surprisingly perfect. The new road over the Plump Hill exposed in its formation, in 1841, an ancient mine hole, in which was found a heap of half-consumed embers, and the skull of what appeared, from its tusks, to be that of a wild boar; the remains, perhaps, of a feast given by our Forest ancestors. Similar vestiges have been met with in other spots.

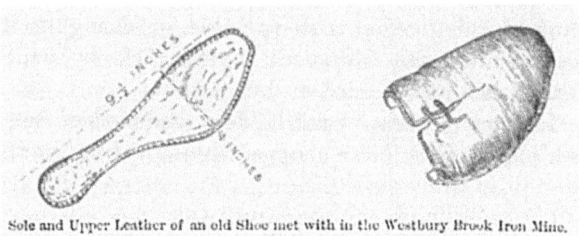

Sole and Upper Leather of an old Shoe met with in the Westbury Brook Iron Mine.

Iron Making in the Olden Times

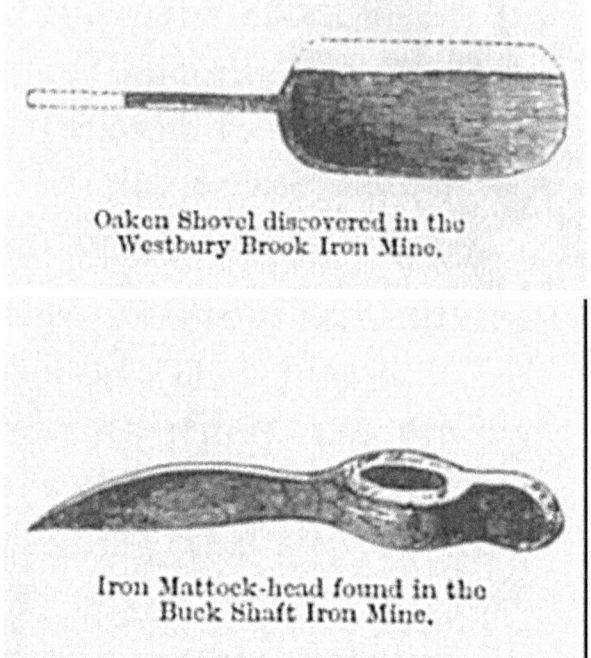

Oaken Shovel discovered in the Westbury Brook Iron Mine.

Iron Mattock-head found in the Buck Shaft Iron Mine.

The earliest historical allusions to these underground works is made by Camden, who records that a gigantic skeleton was found in a cave on the Great Doward Hill, now called "King Arthur's Hall," being evidently the entrance to an ancient iron mine. The next refers to the period of the great rebellion, when the terrified inhabitants of the district are said to have fled to them for safety, when pursued by the troops with which the Forest was infested.

King Arthur's Hall on the Great Doward.

But, whilst no previous mention of these caverns is to be found, nor dates anywhere inscribed on their rocky walls, a clue, as to when and by whom they were first wrought, is given in connection with their metallic products, that abound near them in the state of iron cinders. Thus it is recorded by Mr. Wyrrall, in his MS. description of this subject, that—

> "Coins, fibulæ, and other things, known to be in use with the Romans, have been frequently found in the beds of cinders at certain places. This has occurred particularly at the village of Whitchurch, between Ross and Monmouth, where large stacks of cinders have been found, some of them eight or ten feet under the surface, and demonstrating, without other proof, that they must have lain there for a great number of ages. The writer had opportunities of seeing many of these coins and fibulæ, &c., which have been picked up by the workmen in getting the cinders, in his time; but especially one coin of Trajan, which he remembers was surprisingly perfect, considering the length of time it must have been in the ground. Another instance occurs to his recollection of a

> little image of brass, about four inches long, which was then found in the cinders in the same place, being a very elegant female figure in a dancing attitude, and evidently antique by the drapery."

Numerous other Roman vestiges, on every side of the Forest, may be adverted to. No great distance from Whitchurch, and immediately adjoining this neighbourhood on the north, is the site of Ariconium, marked by numerous traces of the hardware manufacture of that people. Near Lydney and Tidenham, discoveries of Roman relics have been extensively made. At Lydbrook, and on the Coppet Wood Hill, at Perry Grove, and Crabtree Hill, all within or near the Forest—the last being situated in the middle of it—many coins of Philip, Gallienus, Victorinus, and of Claudius Gothicus, have been brought to light. We possess indisputable testimony, from Mr. Lower's researches in the old iron-making parts of Sussex, that the Romans there carried on metallurgical operations at an early period, and we may claim a like antiquity for our Dean Forest workings. An examination of the cinder-heaps that still occur, especially in the precincts of the mines already described, reveals, beyond doubt, the antecedents of the mineral operations of the neighbourhood.

Considering the *extent* of the excavations from whence these metallic relics were procured, it is not surprising that these mounds of slag continue to be constantly met with. Two hundred years ago, they were of course much more abundant, having formed since that period a large part of the supply to the iron furnaces of this district. They are yet numerous enough to catch the eye wherever the observer may direct his steps, either along the retired lane, or in the secluded valley. The fields and orchards, gardens and precincts of the Forest villages, are nearly sure to contain them. Two localities, viz. Cinderford and Cinderhill, no doubt derive their names from them. In some places they have proved so abundant as to have enhanced the value of the land containing them. They even occur on elevated spots, exposed to every wind, and remote from human habitations. Nor is their existence to be limited to the Forest, or even to the Gloucestershire side of the Wye, since the Rev. T. W. Webb states—

> "In many parts of the district of Irchinfield, in Herefordshire, cinders are found in the road, or dispersed in the fields; and in many places heaps of them have been discovered. I would particularly specify the parishes of Tretire with Michaelchurch, Peterstow, great heaps at St.

Weonards, and Llangarren. In the last century, enormous heaps were found at considerable depth in 'the Brook End,' a street of Ross. Near Rudhall, the roads were repaired with them."

Their *character* is peculiar, exhibiting by no means complete fusion, but rather semi-vitrifaction by roasting; the ore retaining, not unfrequently, a large measure of its original weight and form, explained, as Dr. Percy kindly informs me, by charcoal being the fuel employed, and not necessarily arising from want of skill in the operatives. They are said to vary in richness according as they belong to an earlier or later period—so much so, that some persons have ventured, on this data, to specify their respective ages; but other causes may have produced this difference. They exhibit, however, some slight variation of character, indicative, it may be—for so Mr. Wyrrall considered—of relative age, according as they are found to have left in them less or more of the metallic element.

It is impossible to mistake them for common cinders; nor do they resemble the slag of the modern smelting furnace. In fact, they are *sui generis*, and can only be met with where the manufacture of iron was anciently carried on.

Though the constant occurrence of wood embers with the old cinders is conclusive proof that charcoal was the fuel invariably used, yet how it was employed can hardly be determined with exactness, except from what is known of the elementary plans in early use amongst other people, the Egyptians, for instance, the natives of Central Africa, or the iron-workers of Madagascar. A strong draught must necessarily have been made to pass through the ignited fuel, either by placing the furnace so as to take the wind, or by forming it on the principle of the modern wind furnace. Or the required blast might have been created by means of wooden cylinders, or leathern bags, alternately compressed by the hands or feet. Water-power was rarely, if ever, resorted to at this remote date, since cinders are seldom found near brooks or streams.

In common with everything else relating to the manufactures of the kingdom, Domesday Book is silent respecting the mines, iron works, and miners of the Forest. Adverting, however, to this otherwise invaluable return, we learn from it that Edward the Confessor was accustomed to demand from the citizens of Gloucester, "thirty-six dicres of iron, and a hundred elongated iron rods for bolts for the king's ships,"—(xxxvi. dicras ferri & c. virgas ferreas ductiles ad clavos navium Regis). The nearest, and indeed, the only locality, within a distance of many miles, from whence the forgemen of Gloucester could have obtained their iron, was this neighbourhood. Hence the metal they used came from the Forest.

Less than a hundred years later, and all doubt on this point is removed by a notification in the Great Rolls of the Pipe, that 16s. worth of iron was sent, in 1158, to Wudestock (Woodstock) by the king's order, besides 8s. worth more for repairs at his palace. An observation of Geraldus, describing the tour he made through Wales in 1188, speaks of the "noble Forest of Dean, by which Gloucester was amply supplied with iron and venison." [1]

The first charter granted to the Abbey of Flaxley, by Henry II., whilst Duke of Normandy, and therefore previous to 1154, in which year he came to the throne, specifies an iron work at Edlaud, now Elton, near Westbury, on the eastern side of the Forest. [2] His second charter, when king, is more explicit, and describes "an iron forge, free and quit, with as free liberty to work as any of his forges in demesne," showing that he possessed several. The allowance of two oaks per week, wherewith the monks might feed their forge, although not mentioned until 42 Henry III. (1258), when they were commuted for the tract of land yet called the Abbot's Woods, were granted most likely at this period, and afford some data for determining the capacity of the Flaxley works.

At the commencement of his reign (1216), Henry III. commanded "John de Monmouth to cause Richard de Eston to have his forge working in the Forest of Dean, at Staunton, according to the Charter of Henry II." [3]

In the same year, "the Constable of St. Briavell is ordered to remove, without delay, all forges from the Forest of Dean, except the King's demesne forges, which belong to the Castle of St. Briavell, and ought to be sustained with trunks and old trees wherever they are found in the demesnes in the Forest—excepting two forges belonging to Ralph Avenell, concerning which he has the charter of King John, and excepting four 'Blissahiis;' Will. de Dene, & Robert de Alba Mara, & Will. de Abbenhale, & Thomas de Blakencia, and excepting the

---

[1] Hoare's Itinerary of Abp. Baldwin, vol. i. p. 102.
[2] Rudder's Appendix. pp.25, 26.
[3] Rotuli Litterarum Clausarum.

forges of our servants of St. Briavells, which ought to be sustained with dry and dead wood." [1]

Under date 4 Henry III. (1220), "John de Monmouth is commanded not to permit any forge to work, either with green or dry wood, in the Forest of Dean, besides the demesne forges; and to let all those know who have had forges, and who claim to have them by charter or letters patent of our (the king's) ancestors, or our special precepts, that they are to come without delay before H. de Burg, our justiciary, and our counsel, with those letters and charters, that it may be known who may have forges and who may not." [2]

The inference to be drawn from such prohibitory investigations is, that, owing to the remunerative character of the Forest iron works, they had become undesirably numerous, causing an inexpedient waste of the adjoining woods, besides hampering the rights of the Crown.

An immediate effect ensued, as the following memoranda show:—

In the same year as aforesaid, "John de Monmouth is commanded to permit the Abbot and Monks of Flaxlegh to have their forge working in the Forest of Dean, according to the charter of Richard I. (which they have thereof), in the same manner as they had it in the time of King John, notwithstanding that all forges are prohibited in the Forest except the demesne forges." [3]

In the same year, John de Monmouth is commanded "to permit Walter de Lacy to have his forge (fabrica) in the Forest of Dean as he was accustomed to have it, temp. Hen. II. and John, which was prohibited at the time of our general prohibition." Now, also, John de Monmouth received the king's directions as follows:—"William Fitz-Warren has shown the king that whereas Walter de Lacy gave him a forge, which the said Walter and his ancestors have been accustomed to have, temp. Hen. II., Ric. I., and John, and which was prohibited in our general prohibition—we command you to allow the said William to have the said forge (fabrica) moveable in the Forest; but that the forge which the said Walter erected without our order shall remain quiet (remanenta otiosa)."

The next year, 1221, John de Monmouth is ordered to allow Philip de Bantun, Rob. de Alba Mara, John de Lacy, Will. de Dene, Will de Abbenhale, and Thomas de Blakeney, foresters of fee in the Forest of

---

[1] Ibid.
[2] Rotuli Litterarum Clausarum.
[3] Ibid.

Dean, and Nigell Hathway, Martin de la Boze, John Fitz-Hugh, Richard Wither, Rob. Fitz-Warren, Will. Cadel, John Blund, Alexander de Staurs, Roger Wither, John Fitz-Gadway, serventes de feods, to have their "forgias itinerantes ad mortuum et siccum" as they were accustomed to have them temp. Ric. I. and John.[1] A similar privilege was granted, the same year, to Matilda de Cautilupe and Henry, Earl of Warwick—the latter at Lidenie—to have their "forgia," as well as to Walter de Aure to have his "forgia itinerans," and Richd. de Estun his "fabrica."[2] So, likewise, in 1223 (7 Hen. III.), the Monks of Flaxlegh were directed to have "forgiam suam," as in the time of King John.

A document [3] without a date, but unquestionably belonging to the early part of the reign of Hen. III., to whom it seems to be addressed by way of an official report on the state of the Forest, affords the earliest compendium that has been discovered of the extent of its iron works at this period.

Concerning the "Fabricis," these authorities say, that the Monks of Flaxley have "unam fabricam arrantem" at Ardland, in the Forest of our lord the King, and have, where they please, each week, two oaks, &c. &c.

Mabilia de Cautelup has one "fabricam arrantem," at Ettelaw, and three "fossatas" of green wood and one oak for the same, &c. They say, also, that John Malemort (gruyer) holds one "fabricam," &c., and fells one oak each week, where he pleases.

They likewise say that the constable of St. Briavel's Castle holds, in the same place, "unam fabricam," which is sustained by what is felled for the "fabricam" of the said John, and by other perquisites, &c.

Concerning the "Fabricis" which are situated in the vills of the forest. They say that at Bicknour are sometimes four "fabrica," and sometimes two, and sometimes three, from which the Constable takes for each "VII^s. if they be 'arrantes continue' for one year; and the forester, who is forsooth lord of each vill, receives III^d. any way per week from each fabrica; and they are sustained by charcoal made in Wallea, and by perquisites in the Forest." They say, too, that at Ruwardin there are at one time or other "V. fabrica arrantes," and sometimes less, in the same way as the constable and lord of each vill take, as aforesaid.

---

[1] Rotuli Litterarum Clausarum.
[2] Ibid.
[3] Exchequer Department, Forest Rolls, No. 418.

They say also that at Magnam Dene are "VIII^{to} fabricæ arrantes" of charcoal, made without the Forest bounds; and the constable and lord of Dene and of Abbenhal take of the above-named "fabricæ" as is first of all stated.

They say also that at Parvam Dene are "IIII$^{or.}$ fabricæ arrantes" of the perquisites in the Forest, and sometimes also of charcoal made beyond the Forest bounds, and from thence the constable and lord of the same will take as has been already said.

They say also that Nigel of Lideneye holds one "fabricam," at Lideneye, "arrantem" sometimes from the forest, and sometimes with charcoal made in Wallea, from whence the constable receives VII$^{s.}$ per annum.

They say, too, that Walter de Ewies holds one "fabricam arrantem at Lideneye, from whence in the same manner the constable takes as has been said before."

Peculiar interest seems to attach to the above return, not only from its high antiquity, but also because it gives other instances besides that of the monks of Flaxley, in which oak trees were granted to individuals for sustaining their forges. The wording of the report likewise indicates a new meaning, and, no doubt, the correct one—of the term "arrantes," as applied to the forges, which it proves to signify *fed, supported, replenished*, &c., and not *moveable*, as has been heretofore supposed—a term that seems singularly appropriate, considering how rapidly charcoal fuel burns away when urged by a strong blast, and, in consequence, the frequent necessity of renewing it. Besides which, the forge would have to be repeatedly fed with fresh charges of ore.

Gloucester was for ages doubtless the market to which a large portion of the iron made in the Forest of Dean was sent for sale; and so superior was its quality, that Gloucestria, or Glovernia, hardware was much sought after. The following letter—addressed by Simon de Surtiz to Ralph de Wareham, Bishop of Chichester, 1217-1223, or Ralph Neville, who held the see 1223-1245, relative to the purchase of iron, affords an instance of this fact:—

> "I have inspected the letters of his Lordship H. de Kynard, sent to us by you, and which I transmit to your holiness, signifying that he has taken amiss your command respecting the iron to be bought, writing to you that x. marks for the one lot of iron, and c. shilling for the other lot, ought to do.
>
> "Wherefore since the same H. has not rightly understood your command, if it pleases you to write to him,

you will that he have made for you x. marks of inferior iron, supposing he is able to do so. But if otherwise, then that it be v. marks of the heavier and v. marks of the lighter sort, and that the kind made at Gloucester is what is wanted.

"If it please you, write also to the Lord Abbot of Gloucester, so that he work with care for my Lord of Winchester, your guest, that he be as easy as he can about the iron, and without delay.

* * * * *

"Moreover, my Lord H. de Kynard consults you that the iron may be wanted at Bristol and not at Gloucester. But if he yield to your wish I would recommend you that it be brought to Gloucester, as more easy, and without risk. I await your convenience until you can be brought to Winchester."

Amongst the Patent Rolls of the 17th of this same reign (1233), is one entitled "De Forgeis Levand," in Foresta de Dene. And, in 1255, there is another relating to forges in the same. [1]

The issues in money to the Crown, from the mining and making of iron in the Forest, were stated by James Treysil, Custos of the Castle and Manor of St. Briavels, to have amounted to the following sums for the year commencing 13 Jan., 39 Hen. III. (1255), and ending 16 Nov., 40 Hen. III. (1256):—

£	s.	d.	
22	10	0	from the king's great forge, placed out at farm for the time aforesaid.
8	4	6	from foreign forges (*forinsecariis*, *i.e.*, beyond the limits of the Forest, for the same period).
4	9	3	from forges within the Forest.
23	1	4	from the great and little mines.
58	5	1	

---

[1] Inquisition of 15 Edw. III., Exchequer Records, No. 75.

The king's "great forge," mentioned above, yielded, in after years, £26 19s. 3d. to the Custos, but was ere long suppressed, as detrimental to the Forest woods. Its being named here suggests a solution of the term "levantis," or small, generally given to the other forges of the district. They were urged, probably, with such bellows as may be seen carved on an ancient tombstone in Newland Churchyard.

In the year 1841, when that part of the old road leading up to the Hawthorns from Hownal was altered, near the brook, below Rudge Farm, the hearths of five small forges, cut out of the sandstone rock, and curiously pitched round the bottom with small pebbles, were laid open. An iron tube, seven or eight inches long, and one inch and a half bore, apparently the nozzle of a pair of bellows, was also found; as well as scores of old tobacco-pipes, as they seemed, bits of iron, much rusted, and broken earthenware, besides a piece of silver coin. Unfortunately, none of these articles were preserved, or they might have thrown some light on the subject.

The Fabric Rolls of Westminster Abbey, under date 37 Hen. III. (1253), contain the ensuing items:—

	s.	d.
To Henry de Pont, for iron bolts	14	0
To Richard de Celir, for four hundred iron clamps from Gloucester (4 lbs.)	21	0
For carriage of the same iron	3	4
Also—		
To Richard de Celir, for rods of iron from Gloucester (10 lbs.)	16	0
For carriage of the same	6	8

Thus widely spread was the good fame of the Forest metal.

By an Inquisition of the 52 Hen. III. (1268), to ascertain what privileges the abbot and convent of Tynterne were accustomed to have in the Forest, the jurors returned that—"the said abbot and convent, by charters of the King's predecessors, are accustomed to have mines in the Forest for their own forge freely, and have never given anything for the said mines." [1]

---

[1] Exchequer Records, No. 29, Chapter House.

They reported, also, that "by charters of the Earl of Hereford it was granted to the said abbot and convent to have another forge in the said Forest, which was in use in the time of the said jurors."

Ralph de Sandwico, Custos of the Castle and Manor of St. Briavels, in his return of monies received on behalf of the Crown from the iron mines and forges during the 4th of Edward I. (1276), [1] states as paid:—

£	s.	d.	
23	6	9½	from the great and little mines of iron and coal.
11	6	0	for rent of forges in the Forest.
5	15	0	by sale of cinders (cineribus).

This last item seems to show that even then it was customary to use the old cinders left from the still more ancient workings.

A regard of the Forest, [2] taken the 10th Edward I. (A.D. 1282), "De Forgeis in Forestâ," records:—

> "It has been presented by the regarders that many 'forgiæ errantes' have been and are still in the Forest, and that those who have held and still hold them commit many evils in the Forest, above the wood and beneath it, both by injuring the trees as well as by means of their forges, great detriment being done in the Forest by them and their wood colliers. And these are the names of such as have held or still hold them, viz.:—

"*Of Parva Dene.*—Ralph Page, William the sharesmith, Thomas Hewlin, William of Hereford, John of Hereford—in all 5.

"*Of Blakeney.*—Hugo Textor, together with Walter of Blakeney, Adam of Erlyxforde (Ayleford), John Boyce—altogether 3.

"*Of Ettelano* (Etloe).—Richard Pole.

"*Of Lideneye.*—John Scot, Stephen Edys.

"*Of East Sancto Briavello.*—Roger Spore, Adam Betrech, Stephen Marlemort, Nicholas the Pichehere, John Hurel, Philipp Martin, Henry the Bole, Adam Fitawe, Richard Walensis, John Missor, Henry Fitz William of

---

[1] Inquisition of 15 Edw. III., Exchequer Records, No. 75.
[2] Exchequer Records, Chap. V. f. 18, No. 18, Col. I.

Tullic, William the jailer, William of Ruerdean—altogether 13.

"*Of Staunton*.—Robert le Noreys, Godfrey le Stempore, William Dorby, Nigel de Staunton, Adam le Coliere, Thomas the jailer, William Cambel, Peter le Monner, Philipp the clerk, William Clayneberd—in all 10.

"*Of Bikemore* (English Bicknor).—Walter Pisum, another by the same.

"*Of Hopo Malhoysel* (Hope Mansel).—The Abbot of Gloucester, Henry Duke of Gloucester.

"*Of Reruwardin*.—William, son of Matilda, Roger Fowel, Nicholas Charlemayne, Thomas Mone, Roger Kingessone, Thomas le Leye, William Baret, William Jordan—altogether 8.

"*Of Magna Dene.*—Adam Simund, Robert le Paumer, Reginald Balloc, Hugo le Paumer, Robert de la Zone, Galfrid the Nailer, Robert Dun, Thomas Balloc, Hugo Godwyn, Phelicia Pecoe, John Geffrey, Nicholas Drayclasz, Galfrid Dobel, Richard Strongbowe—in all 14."

According to this return, there were 60 forges (fabrica) at work in the district of the Forest at this period. Of these, 19 were situated on the east side, towards Gloucester; 6 were on the south, towards the Severn; 23 were on the west, towards the Wye; and 12 were on the north, towards Herefordshire. Hence, they were most numerous on the east and west, especially the latter, in accordance with the greater extent of the ancient mine holes on that side.

The annual charge levied by the Crown for each forge was usually at the rate of 13s. 6d., or a mark. When otherwise—for in certain cases it amounted to 20s.—it arose, probably, from some local circumstances connected with the quantity or quality of the iron made at that particular work.

Taken altogether, the forges in the Forest now yielded the king more than £30 every twelve months.

They were leased for periods varying from three months to twenty years, although the general length approached much nearer to the shorter limit than to the longer one.

By the same "regard," the iron mines are specified as follows:—

"The jurors say that Ralph de Abenhalo hath a mine in the bailiwick of Abenhale. And our Lord the King hath nothing except six semes (eight bushels) of mine ore each week, and giveth for it to the work people VI^d.

"The Church of Michegros hath a mine in the bailiwick of Bikenore if it should be found (inventa). Walter de Astune claims a mine in the bailiwick of Blackeneye, if it should be found.

"Our Lord the King hath a mine in the bailiwick of Magna Dene, and he takes from each workman who shall gain every three days three semes of mine ore, 1^d. per week. And when a mine is first of all found, our Lord the King shall have one man working with the other workman in the mine, and hire him for 2^d. a day, and he shall have such profit as he may find by the one workman. Item, our Lord the King shall have from thence each week, six semes of mine ore, which is called 'Lawe ore.' And he shall give for this to the workmen VI^d. a week. [1]

"Our Lord the King hath in the bailiwick of the Birs, because there are there more mines than in the bailiwick of Dene, all as if in the bailiwick of Dene, this excepted, that he hath from thence each week XXIV. semes of mine, which is called 'Lawe ore.' And he giveth for this to the workman, every seven days, 11^s."

"Our Lord the King hath in the bailiwick of Staunton a mine, and he takes from thence, all as if in the bailiwick of Magna Dene, this excepted, that our Lord the King shall have for each workman that gains each week three semes of mine ore, ½^d. every seven days and not more.

"Item, if our Lord the King shall have a 'forgeam arrantem,' the aforesaid workmen shall bring him mine ore for the supply of the aforesaid forge. And our Lord the King shall give them for each seme 1^d.

---

[1] "Dominus Rex habet mineriam in Balliva de Magnâ Dene. Et capit do quolibet operaris qui poterit lucrari per septimanam tres summas minea 1. denarium per septimanam. Et quando minea primo invenietur Dominus Rex habebit unum hominem operantem cum aliis operantibus in mineria, et conducet illum pro duobus denariis per diem, et habebit partem lucri quantum eveniat uni operaris. Item, Dominus Rex habebit unde per septimanam sex summas mineæ quæ vocantur 'Lawe ore.' Et dabit propter hoc operariis VI. denariis per septimanam."

"Item, our Lord the King shall have for each seme of mine ore that is taken out of the Forest, $\frac{1}{2}^d$.

"And all that our Lord the King takes from the mine are put to farm for £46.

"Item, in the bailiwick de Lacu is a mine, and our Lord Richard Talebat holds it, and it is unknown by what warrant. And our Lord the King takes nothing from it.

"Item, the Earl of Warwychiæ hath a mine in his own wood of Lideneye, and our Lord the King takes nothing from it, except for the mine which is carried out of the Forest, a $\frac{1}{2}^d$. The jurors say that the foresters take cooper's stuff out of the open woods from the miners to the inbondage of the miners, and work it for their own profit."

From the above curious items it appears that the iron mines, in common with the forges, were mostly situated on the Wye side of the Forest. But then the bailiwicks of Little Dean and Ruerdean are not included.

It would appear, too, that the ore was then measured by the bushel, as it has been ever since, owing, of course, to its loose powdery nature, which seems, therefore, to have been the sort preferred.

The other singular particulars descriptive of "lawe ore," &c., are noticed elsewhere, in the second portion of this work, containing the "Book of Dennis."

Another "Inquisition" exists, bearing date the 12th Edward II.,[1] but applying to the year commencing with Easter the 10th Edward II., or thirty-five years later than the former return. It was made at (Mitchel) Dene, on the Wednesday before the feast of St. Nicholas (6th December), by Lord Ralph de Abbendale and other foresters of fee, and by twelve jurors.

It assigns one "fabrica," consuming ten shillings' worth of wood-coal per week, or £24 yearly, to each of the following persons, located as under:—

"*At St. Briavel's.*—Nicholas Le Prichure (who makes ploughs), Philip Hurel (making 'grossum ferrum'), Richard Walencius, William FitzOsbert, Adam Betricz, Roger

---

[1] Chapter House Records.

Spore, John Le Hayward, Stephen Malemort, William Bocod—in all 9.

"*At Stanton.*—Philip Clerk, Thomas Jan,—Walding—total 3.

"*At Ruwardyn.*—Roger Fowel, Peter de Obre, William Buysche, John Kole, Celimon Le Dine, with William Le Smale, William FitzMaud, Thomas de Leye, Adam de Leye (making ploughs), Robert Smart, Peter de Huwale, Walter de Wyteling, Thomas de Leye—in all 12.

"*At (Mitchel) Dene.*—Galfridus Dobel, Nicholas Draylax, John Geffray, Richard Stranglebowe, Richard de Gorstleye, Hugo Godewyne, Robert Down, Robert, son of Roger de Ponte, Hugo le Powmer, Margary de la Lond, Reginald Rouge, Robert Palmer, Thomas Bulloc—in all 13.

"*Parva Dene.*—John Hereford, Thomas Lewelin—total 8.

"*Erleyeforde*, i.e. Ayleford.—Adam de Erleyeforde, Robert Pote, Stephen Edy, John Schotticus—altogether 4."

If this list includes all the forges then at work in the Forest, a diminution of seventeen had occurred during the last thirty-five years, and apparently on the west side of the district. Changes may also be observed to have taken place in the owners, although several names are met with a second time.

Considerable prosperity and steadiness continued to attend the mining and making iron in the Forest, so as to render it eligible for the imposition of tithes. So, on the completion of Newland Church, at this period, the Bishop of Llandaff, who presented to it, applied for and obtained from Edward III., in the fourteenth year of his reign, A.D. 1341, a grant of the tenth part of the ore raised in the neighbourhood, which, together with the forest forges, yielded a rental of £34 the same year.[1]

To the Edwardian period, that has now, by the aid of the numerous records already quoted, been so minutely substantiated, must be assigned the most prosperous era of the Forest of Dean iron works. A large portion of such success is to be traced to the celebrity at this date of the great fair in Gloucester. It began annually on the eve of St. John

---

[1] Inquisition of 15 Edward III., Exchequer Records.

Baptist's day, and continued for the five days following. Agricultural implements were in much request at it, and even noblemen are said to have attended. [1]

Other places, such as Caerleon, Newport, Barkley, Monmouth, and Trellech, obtained their supplies of iron, or at least the mine-ore, from this neighbourhood, the Forest miner having a certain status of his own, and constituting, with his partners or "verns," a guild of considerable local influence. [2]

The heraldic crest (p. 67) forming part of a mutilated brass of the fifteenth century, within the Clearwell Chapel of Newland Church, gives a graphic representation of the iron miner equipped for his work, if not actually engaged in it. He is represented as wearing a cap, and holding between his teeth a candle-*stick*, an appurtenance still in use amongst the miners about Coleford, as may be observed by examining the frontispiece to this volume, thus illustrating the primitive use and significance of the phrase candle-*stick*. With the small mattock in his right hand, he would loosen the fine mineral earth lodged in the cavity within which he worked, as occasion required, or else detach the metallic incrustations lining its sides. A light wooden mine hod, covered, probably, with hide, hangs at his back by a shoulder-strap, fastened to his belt. His attire is completed by a thick flannel frock and leathern breeches, tied with thongs below the knee. The feet most likely were bare.

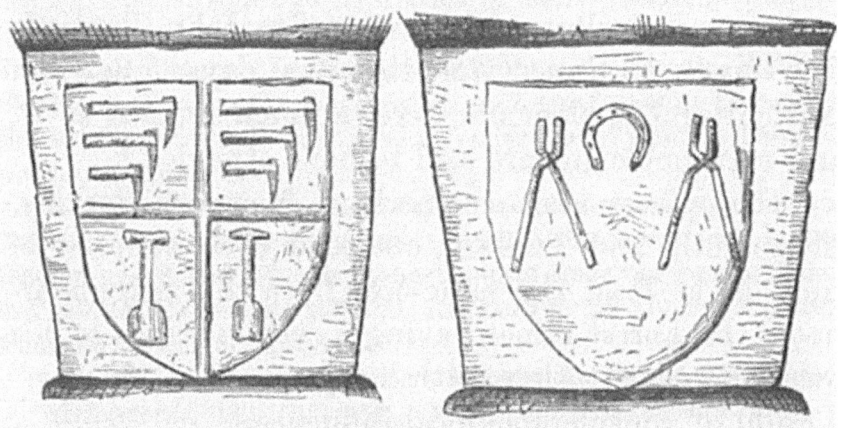

Representation of Miners' and Smiths' Tools, on the Font, in Abbenhall Church, A.D. 1450.

---

[1] Fosbrooke's Gloucester, p. 44.
[2] Book of Mine Law.

Iron Making in the Olden Times

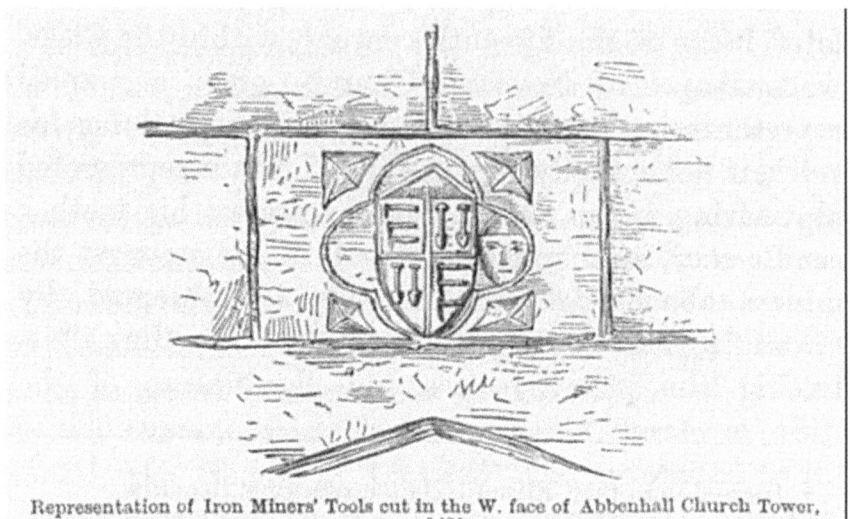

Representation of Iron Miners' Tools cut in the W. face of Abbenhall Church Tower, A.D. 1450.

Other contemporary representations of the mining implements in use at this time in the Forest occur at Abbenhall, where the west side of the church tower, and also the font, exhibit panels carved with hammers, shovels, &c.

Some persons of considerable experience have concluded that the ore was washed ere placed in the forge. The mounds of deep red earth that occur in some parts of the Forest are supposed to establish this practice. If ever such was the custom, no other trace of it appears, and it is quite unknown now. In parts of South Wales, water may be used with advantage, but were it applied to the mineral here, much would be washed away, because of its finely divided state.

An interval of two centuries and upwards intervenes at this point. No data for determining the state of the Dean Forest iron works again occurs until the reign of Elizabeth. For the mean time, however, there seems every probability that operations went on without intermission, although some decline had apparently taken place. Perhaps the dissolution of the monasteries interrupted the works at Flaxley and Tintern, by causing the discharge of the old hands and the employment of unskilled operatives in their stead.

The domestic series of the State Papers enable the clue to be resumed under date 30th June, 1566, when one William Humfrey, upon information derived from some German miners, addressed a letter "to Sir Wm Cecill ab^t the plenty of good iron contained in the Forest of Dean." It was, no doubt, the general rumour of this fact that rendered it an object of spoliation to the would-be invaders from Spain in

1588. At this date, wire, drawn by strength of hand, is said to have been made at Sowdley. For such kind of manufacture the Forest iron, from its toughness and ductility, was admirably fitted, without requiring any essential change in the mode of reducing the ore, although improved methods of doing so were being adopted in other parts of the kingdom, particularly in Sussex. That the old way of working lingered long in the northern counties appears from a statement of Mr. Wyrrall's, to the effect that "The father of the late Mr. James Cockshut of Pontypool found, some years ago, an old man working by himself at a bloomary forge in a remote part of Yorkshire. Being himself well acquainted with every branch of the iron trade and works, he stayed with the man long enough to investigate and comprehend his mode of working, and saw him work, with his own hands, a piece of iron from the ore to the bar."

The earliest intimation of any decided alteration being adopted in the manner of operating on the raw metal occurs in the terms of a "bargayne" made by the Crown "wth Giles Brudges and others," [1] on 14th June, 1611, demising "libertye to erect all manner of workes, iron or other, by lande or water, excepting wyer workes, and the same to pull downe, remove, and alter att pleasure," with "libertye to take myne oare and synders, either to be used att the workes or otherwise," &c. By "synders" is meant the refuse of the old forges, but which, by the new process, could be made to yield a profitable per centage of metal, which the former method had failed to extract.

Early in the year following (17 Feb. 1612), a similar "bargayne" was made with no less a person than William, Earl of Pembroke, elder brother of Sir Philip Herbert, one of James I.'s earliest favourites. His high position did not prevent him, therefore, from engaging in manufacture and trade, only in the prosecution of them he would be made to pay accordingly. Thus, whilst the former party paid 3s. for each cord of wood, the earl was charged 4s. for 12,000 cords yearly for twenty-one years, or £200 per annum, with £33 6s. 8d. besides, all for fuel only. He was, however, "to have allowance of reasonable fireboote for the workmen out of the dead and dry wood, and to inclose a garden not exceedinge halfe an acre to every howse, and likewise to inclose for the necessity of the workes the number of XII. acres to every severall worke; the howses and enclosures to be pulled downe and layd open as the workes shall cease or remove."

---

[1] Lansdowne MSS., No. 166, f. 365.

Iron Making in the Olden Times

Similar appreciation of the remunerative character of iron making occurs in connection with a still more illustrious person. There exists a letter, dated 7 May, 1611, addressed by Sir Francis Bacon to Cecil, Lord Salisbury, endorsed, "Ld Lisle, Sir F. Bacon, and others, to be preferred in the sale intended in the Forest of Deane for some reasonable portion of wood, for maintenance of their Wire-works, paying as any others."

The letter itself runs in these words:—

"It may please your good Lordship,

"Understanding that his Majesty will be pleased to sell some good portion of wood in the Forest of Deane, which lies very convenient to the Company's Wire Works at Tynterne and Whitebrooke, we are enforced to have recourse to your lordship, as to our Governor of the said Company, humbly praying your lordship to afford us some reasonable quantity thereof, the better to uphold the said works, whereof by information from our farmers there, we stand in such need, as without your lordship's favour we shall hardly be able to subsist any long time. We do not entreat your lordship for any other or more easy price than that your lordship directs the sale of it to others; only we humbly pray for some preferment in the opportunity of the place where the woods lie, and in the quantity, as it may answer in some portion to our wants. Herein, if your lordship will be pleased to favour us, then we humbly pray your lordship to direct us to some such persons as your Lordship resolves to employ in the business. And as we humbly take our leaves of your lordship,

"Your lordship's humbly at command.

"London."

What success attended this application, or the enterprise which it was intended to promote, does not appear. Wealth flowed in from other quarters, so that the great philosopher was relieved from the necessity of trying to make money by making iron. Tyntern, however, and also Whitebrook, have ever since been connected with that kind of manufacture.

A third "bargayne," and corresponding with the two previous ones, was agreed to on the 3rd May, 1615, with Sir Basil Brook, from whom rent in kind was thus retained:—"iron, 320 tons p. annum, wch att xii^{ll}

x^s the tonn, cometh to 4000 per an.: the rent reserved to be pay^d in iron by 40 tonns p. month, wch cometh to 500^{ll} every month; so in toto yearelye 4000^{ll}."

A proviso was added that—"The workes already buylt, onlye grantted wth no power to remove them, but bound to mayntayne and leave them in good case and repayre, wth all stock of hammers, anvils, and other necessarys received att the pattentees' intyre," as also that "libertye for myne and synders for supplying of the workes onlye, to be taken by delivery of the miners att the price agreed uppon."

Great confidence was reposed in Sir Basil Brook, since he, with Robert Chaldecott, obtained a contemporary grant of the office of clerk or overseer of the iron works in the Forest for fifteen years. [1] But so much did they abuse it, that ere three years had elapsed, a commission was issued, 17 July, 1618, to Sir Thos. Brudnell, Sir John Tracy, Sir William Cooke, and others, [2] "to survey and examine the wastes made in the Forest of Dean by Sir Basil Brooke and others, farmers of iron works there." In their report, one item states that "His Majestie, since the erecting the iron works, had received a greater revenue than formerly." They were to proceed on interrogatories prepared by Sir Wm. Throgmorton, Bart., who was himself engaged in the like manufacture, [3] being associated therein with Sir Sackville Crowe, Bart., John Taylor, and John Guernsey, of Bristol, merchant farmers of his Majesty's iron works. Sir Edward and Sir John Winter, of Lydney, and Henry, Lord Herbert of Ragland, had iron works as well.

In April, 1621, [4] Messrs. Richd. Challoner and Phil. Harris, tenants to Lord Robartes, appear to have succeeded to the works formerly held by Sir Basil Brook. Within four years, however, one Christ. Bainbridge obtained judgment against them for cutting down 1200 trees for their own purposes, but they were ultimately pardoned, as likewise their predecessors, who had become liable for £33,675 16s. 8d.

The name of Sir Edwd. Villiers now appears [5] as renting iron works in the Forest; then that of Sir Richd. Catchmay, having Wm. Rowles and Robt. Treswell for his overseers.

---

[1] State Papers.
[2] Ibid.
[3] Ibid.
[4] Ibid.
[5] State Papers.

Amidst these successive changes, the only person who seems to have continued in uninterrupted possession of his works for making iron, was William Earl of Pembroke, Lord Steward. In 1627 he had the lease of them renewed to him for twenty-one years. By him, probably, the 610 guns were cast, as ordered by the Crown for the States General of Holland, A.D. 1629. The spot where they were made was, it would seem, ever after called "Guns Mills," and by which name it is still known. Guns Pill, on the Severn, was the place, doubtless, where they were afterwards shipped.

An inventory, unique, probably, in its singularly explicit description of the buildings and machinery used by the above-named manufacturers, and bearing the date of 1635, happily came under Mr. Wyrrall's observation, and was by him carefully transcribed. We learn from it that the stone body of the *furnace* now used in the neighbourhood was usually about 22 feet square, the blast being kept up by a water-wheel not less than 22 feet in diameter, acting upon two pairs of bellows, measuring 18 feet by 4, and kept in blast for several months together. Such structures existed at Cannop, Park End, Sowdley, and Lydbrook. Besides which, there were *forges*, comprising chafferies and fineries, at Park End, Whitecroft, Sowdley, and Lydbrook.

A SURVEY OF THE FOREST OF DEAN IRON WORKS IN 1635.

"*Canop Furnace.*—Most p^t new built, the rest repaired by the Farmers, 22^{ft} square, wheel 22^{ft} diam^r. Furnace box built years since by the Farmers. Bridge-house 48^{ft} by 21, 9 high, built 4 years. Bellows boards 18^{ft} by 4. Clerk's house and stable built by the Farmers. A cottage built by the Workmen belonging to the Works, now occupied by the Filler. Built before the Farmers hired. Founder's house, Minecracker's cabin, a Mine Kiln.

"*Park Furnace.*—Same dimensions, repaired 4 years since by the Farmers, Wheel and almost all the houses built by the Farmers.

"*Park End Forge.*—2 Ham^{rs}, 3 Fineries, 1 Chaffery, rep^d 2 years since, one of the Fineries new.

"*Whitecroft Forge,*—built ab^t 6 y^{rs} since by the Farmers, do do

"*Bradley Forge.*—do do do

"*Sowdley Furnace*, built 3 years—Qu. if rebuilt? Bridge house, p^t built by the Farmers, p^t old and decay^d, Trow leading to the wheel, ½ made new 5 years since, decay^d, 5 Cottages, 1 built by the Farmers. A dam a mile above Sowdley built by the Farmers. A dam half a mile still higher, built long since.

"*Sowdley Forge*, 2 Fineries, 1 Chaffery built 2 years, in the place of the old Forge. Trows and Penstocks made new by the Farmers, decayed.

"*Lydbrook Furnace*, 23^{ft} long, 9 bottom, 23^{ft} deep, new built 3 y^{rs} since from the ground, 3^{ft} higher than before, much cracked. A great Buttress behind the Furnace to strengthen it.

"*Lydbrook Forge.*—1 Chaffery, 2 Fineries, House built 4 years, being burnt by accident."

Besides the above, Mr. Wyrrall transcribed the following additional particulars from a MS. dated 23 September, 1635, and endorsed,—

Iron Making in the Olden Times

"The booke of Survey for the Forest of Deane Iron work, and the Warrant annexed unto y^t."

"*Cannope Furnace.*—Now blowinge, and likely to contynue aboute 3 weeks. The most part new built, and the rest repaired by the Farmers about 4 years since. Stone walls, about 60^{lb}, consistinge of the stone body thereof 22 foote square, wherein are:—

"In the fore front 4 Sowes of Iron)
) 7 Sowes.
and the Tempiron Wall 3 Sowes)

"A Wheele, 22 Foote diam^r, 7 Iron Whops, one the Waste, made about three years since. With shafte and all things belonging about 20^{lb}, in good repaire.

"The Furnace Howse half tiled, built with timber 4 years since by the Farmers, cost about 80^{lb}, in repaire.

"The Bridge House, 21 foot broad, 48 foot longe, and 9 foote heigh, built about 2 years since, the bridge about 4 years, covered with bords bottomed with Planks.

"5 bellow bords ready sawed, 18^{ft} longe, 4^{ft} broad. A Watter Trowe 1^{ft} at bottome and 15 ynches high, 75 yards long, leadinge the water to the Wheele, cut out of the whole tymber, and ledged at the top, newe made within 4 years, and now in repaire, cost about 20^{lb}.

"The Hutch leading the Watter from the Wheele, 5 foot square, 85 foote long, not mended by these Farmers, in repaire.

"In doinge of the saied Workes, besides the Hutch used by estimate about 150 Tonns, at VIII^s, and the Hutch about 40 Tonns, being trees only slitt and clapt together at 5^s the Ton.

"*Outhouses.*—The Furnace Keeper's Cabbyne built of timber covered with bords built by the Farmers, cost 3^{lb}, 4 tonns.

"A Cottage neare the said Furnaces built by the workmen of the said Works, now enjoyed by the Filler there, and not belonging to the Workes.

"A Howse wherein the Clarke dwells, built by the Farmers wth a stable, 20 Nob^s 6 Tonns.

"Another howse adjoyninge for the founder, built before the Farmers' time.

"Another little cabbyne for the Myne Cracker, built before the Farmers' time.

"8 dozen of Collyers' Hurdles, 13^s 4^d.

"A Myne Kilne not in repaire, built before the Farmers' tyme, with 5 piggs of Iron in the walls, 20^s will repaire.

"Cole places.

"*Implen^{nts}*—one paire of Bellowes furnished with Iron implem^{nts}, somewhat defective in the lethers, valued at 15^{lb}, made by the Farmers, the repaire whereof will cost 6^{lb} 13^s 4^d.

"6 cambes of iron in Wheele Shaft waying about 4^{cwt}.

"3 water Trowes for the Worke.

"1 Grindstone, 19 longe Ringers, 1 short one, one Constable, 7 Sinder Shovells, 1 moulding Ship, 2 casting ladles, 1 Cinder-hooke, 1 Plackett, 2 buck stoves, 1 Tuiron hooke, 1 Iron Tempe, 1 Sinder plate, 1 dame plate.

"4 Wheele barrowes, 1 great Sledge, 1 Tuiron plate cast, 1 Shamell plate, 1 Gage, 1 crackt wooden beam and scales, furnished, and triangles, 1 ton of W^{tts}, Pigs used for weights upon the bellows poises. 3½^c of Rawe Iron, 1 new firkett in the Backside, 1 lader of 14 rungs, 1 dozen of cole basketts, 2 Myne hammers, 2 Myne Shovells, 2 Coale Rakes, 2 Myne Rakes, 2 baskes to put myne into the Furnace.

"*Parke Furnace.*—The stone body thereof 22 foote square in the Front, 2 broken Sowes, one taken thence, 2 sowes in the Wall.

"Repaired 4 years since by the Farmers, viz., the backe wall from the foundation to the top, and parte of the wall over the Bellows, 40^{lb} it cost.

"The Water Wheele 22 feet heigh, wth a Shaft whereon 7 whops, 2 Gudgions and 2 brasses, built about the same tyme, in repaire, valued at 20^{lb}. The Furnace Howse tiled, built with stone wall 9 foot heigh, 22 foote square, the Roof good, built about the same tyme, in repair, saving a Lace by the Bridge. The stone worke valued at 10^{lb}. The Carpenter's worke one the roof at 20^s, the tilinge valued at 6^{lb} 13^s 4^d.

"A Pent house under the Furnace, 10^s.

"The Bridge House 42^{ft} longe, 22^{ft} broad, the said walles, 8½ foot, covered with boards, double bottomed with plancke, upon stronge sleepers, valued at 40^{lb}.

"Fence Walls all built by the Farmers about 4 yeares since.

"100 foote of trowes made of square timber, hollowed and covered with plancke, valued at 10^{lb}, made by the Farmers.

"Another Water course, built with stone one both sides and covered wth planckes 2½ foot broad, 46 foot, in repaire, 5^{lb}.

"An Iron cast grate one the same watercourse.

"A watercourse of half a mile one the North of the Furnace, at the head thereof a dam and a small breach, wants soweringe, otherwise good, cutt by the Farmers, and cost them 20^{lb}, and will cost 3lb.

"A Water course of above ½ mile to the South, made before their tyme.

Iron Making in the Olden Times

"The Hutch built with stone and covered with plankes of 6 foot heigh, 3 foot broad, 70^{ft}, saving about 11 foot at the vent which is timber, repaired by the Farmers, in repaire, but the Courant stopt below with cinders, 13^{lb} 6^s 8d; the cutting of a newe will cost 8lb.

"The Fownder's howse built before the Farmers' tyme.

"A Cottage adjoininge.

"A Cabbyne for the bridge-server, covered with boards, built by them about a yeare since, 3 tonns, 18^{ft} longe, 11 broad, valued at 5^{lb}.

"A Cabbyne adjoining to the Furnace for the Furnace Keeper, about a Tonn, built by the Farmers, and valued at 2^{lb}.

"A Faire Howse, the ends stone built, the rest with Timber 50 foot longe, 16 broad; in it is a crosse building ? stories heigh, in repaire, tiled, built before the Farmers now granted, with stables belonging, of tymber.

"A smale cottage, now William Wayt's.

"A myne kylne, the inside in decay, the piggs of iron taken out of the draught thereof, the repaire will cost 2lb.

"Tymber in doeinge of)
) 150 tonnes, VI^s VIIId the tonne.
the saied worke)

"*Implem^{nts}*.—1 p^r bellowes open with the furniture of iron thereto belonging, defective in the lethers, valued at 13^{lb} 6s 8d, the repaire will cost 10^{lb}.; 2 buckstaves, 1 dam-plate, 2 cinder plats, 1 tuiron plate, 1 plackett, 1 gadge, 1 tuiron hoocke, 1 dam hoocke or stopinge hoocke, 4 iron shovells, 9 ringers, 6 cole baskets, 2 wheel barrows, 2 myne hammers, 1 coale rake, 2 cinder raks, 1 great sledge, 1 ringer hammer, 1 constable, 1 shammell plate, 6 iron cambs.

"A beame with scales, hoocks, triangles and lincks, with about ½ a ton of rawe iron for a w^t, in repaire: 1 sowe of iron of 16$^{cwt.}$ which was in the front wall, soe now lyes before the doore, 5^{lb}).

"1 Grindstone, 2 bellowe boards, never used, and 4 old ones, 1^{lb} 10$^{s.}$

"Collyers' Hurdles.

"The tymber ymployed about the said worke estimated at 140 tonns, and valued at 8s the tonn, 56^{lb}.

"The Repaire of the body of the furnace and the buildings, beames thereto belonginge, and other defects, to make it fit to blowe, estimated at 60^{lb}.

"*Park End Forge*—consistinge of 2 hamers, 3 Fyneryes and 1 chaffery, repayered about 2 years since by the Farmers, viz., 2 newe drome beames, 2 great hamers, shafts with wheeles and armes all newe, the body of the forge repaired in sundry places, one of the fyneryes built newe with the whole and shafts.

"The harmes to the great hamers newe and in repaire valued at 12^{lb}.

"One other finerye chimney, made within the yeare, 5^{lb}, 3 newe trowes through the bay, 26^{ft} longe a piece, covered with planke one the west side, 13^{lb} 6^s 8^d.

"The hamer hutch one the west side, heigh and broad one the one side, plancked in the bottome ranges of tymber with spreaders conteyninge 150 foote in length, 40^{lb}.

"The chaffery wheele in the west side, old and decayed, 3^{lb} to repaire it.

"One longe trowe one the est side leading the watter to the fynerye, 66 foote longe, 6^{lb} 13^s 4^d; another great trowe with a penstocke, 32 foote, cost 3^{lb} 6^s 8^d; 1 great penstocke in the hamer trowe, 14 foot longe, 2 foote square, 40^s.

"2 Water Pricke Posts with his laces, 4^{lb}.

"The Hamer Hutch one the west side, 4 foote square, bottoms and sides with plancks, 2 ranges of timber 150 foote longe, 10^{lb}.

"The bodye of one Fynerye wheele all newe, made within 2 yeares last past by the Farmers.

"One little house for the carpenter to work in one the bay.

"Two ranges of tymber worke in the lower side of the bay, consistinge of sils, laces, and posts, built by the Farmers within 2 yeares, 120 foote, 12 heigh, 80^{lb}.

"The front of the bay where the water is led to the west side and drawinge gates built about 2 years since. Stone walls on each side, 5^{lb}.

"A flowd gate with 6 sluices, strongly tymbered, built with stronge wall one either side thereof, 160 foote longe, 3^{ft} heigh, 3 foot thicke, aproned and plancked on the top for a bridge 3 years since, 44 foot longe, 22^{ft} broad, 50^{lb}.

* * * * *

The same careful investigator (Mr. Wyrrall) of every particular relating to the iron works of the Forest, formed a glossary of the terms used in the above specifications, which not only sufficiently explains them, but also shows that very similar apparatus continued to be used in this neighbourhood up to the close of the last century. It proceeds thus:—

"*Sows of Iron* are the long pieces of cast iron as they run into the sand immediately from the furnace; thus called from the appearance of this and the shorter pieces which are runned into smaller gutters made in the same sand, from the resemblance they have to a sow lying on her side with her pigs at her dugs. These are for working up in the forges; but it is usual to cast other sows of iron of very great size to lay in the walls of the furnaces as beams to support the great strain of the work.

"*Dam Plate* is a large flat plate of cast iron placed on its edge against the front of the furnace, with a stone cut sloping and placed on the inside. This plate has a notch on the top for the cinder or scruff to run off, and a place at the side to discharge the metal at casting.

"*The Shaft* of a wheel is a large round beam having the wheel fixed near the one end of it, and turning upon gudgeons or centres fixed in the two ends.

"*The Furnace House* I take to be what we call the casting-house, where the metal runs out of the furnace into the sand.

"*The Bridge* is the place where the raw materials are laid down ready to be thrown into the furnace. I conceive that it had its name (which is still continued) from this circumstance—that in the infancy of these works it was built as a bridge, hollow underneath. It was not at first known what strength was required to support the blast of a furnace bellows; and the consequence was that they were often out of repair, and frequently obliged to be built almost entirely new.

"*Bellows Boards*—Not very different from the present dimensions.

"*Water Troughs*—scooped out of the solid timber. This shows the great simplicity of these times, not 150 years ago.

"*The Hutch*, or as it is now corruptly called the Witch, a wide covered drain below the furnace-wheel to carry off the water from it, usually arched, but here only covered with timbers to support the rubbish and earth thrown upon it.

"*Cambs* are iron cogs fixed in the shaft to work the bellows as the wheel turns round.

"*Cinder Shovels*, iron shovels for taking up the cinders into the boxes, both to measure them and to fill the furnace.

"*Moulding Ship*, an iron tool fixed on a wooden handle, so formed as to make the gutters in the sand for casting the pig and sow iron.

"*Casting Ladles*, made hollow like a dish, with a lip to lade up the liquid iron for small castings.

"*Wringers*, large long bars of iron to wring the furnace, that is to clear it of the grosser and least fluid cinder which rises on the upper surface, and would there coagulate and soon prevent the furnace from working aright.

"*Constable*, a bar of very great substance and length, kept always lying by a furnace in readiness for extraordinary purposes in which uncommon strength and purchase was required. I suppose this name to have been given to this tool on account of its superior bulk and power, and in allusion to the Constable of St. Briavel's Castle, an officer heretofore of very great weight and consequence in this forest.

"*Cinder Hook*, a hook of iron for drawing away the scruff or cinder which runs liquid out of the furnace over the dam plate, and soon becomes a solid substance, which must be removed to make room for fresh cinder to run out into its place.

"*Plackett*, a tool contrived as a kind of trowel for smoothing and shaping the clay.

"*Buckstones*, now called Buckstaves, are two thick plates of iron, about 5 or 6 feet long, fixed one on each side of the front of the furnace down to the ground to support the stone work.

"*Iron Tempe* is a plate fixed at the bottom of the front wall of the furnace over the flame between the buckstaves.

"*Tuiron Plate* is a plate of cast iron fixed before the noses of the bellows, and so shaped as to conduct the blast into the body of the furnace.

"*Tuiron Hooke*, a tool contrived for conveying a lump of tempered clay before the point of the tuiron plate, to guard the wall from wearing away as it would otherwise do in that part, there being the greatest force of the fire.

"*Shammel Plate*, a piece of cast iron fixed on a wooden frame, in the shape of a ⌐, which works up and down as a crank, so as for the camb to lay hold of this iron, and thereby press down the bellows.

"*Firketts* are large square pieces of timber laid upon the upper woods of the bellows, to steady it and to work it.

"*Firkett Hooks*, two strong hooks of square wrought iron fixed at the smallest end of the bellows to keep it firm and in its place.

"*Gage*, two rods of iron jointed in the middle, with a ring for the filler to drop the shortest end into the furnace at the top, to know when it is worked down low enough to be charged.

"*Poises*, wooden beams, one over each bellows, fixed upon centres across another very large beam; at the longest end of these poises are open boxes bound with iron, and the little end being fixed with harness to the upper ends of the firketts are thus pressed down, and the bellows with it, by the working of the wheel, while the weight of the poises lifts them up alternately as the wheel goes round."

Iron Making in the Olden Times

* * * * *

As to the length of time these works continued in operation, the late Mr. Mushet, who knew the district intimately, in his valuable papers on iron, &c., considered that they were abandoned shortly after the date of the inventory, *i.e.* 1635, since, "with the exception of the slags, traces of the water-mounds, and the faint lines of the water-courses, not a vestige of any of them remains."

He adds,—

> "About fourteen years ago I first saw the ruins of one of these furnaces, situated below York Lodge, and surrounded by a large heap of slag or scoria that is produced in making pig iron. As the situation of this furnace was remote from roads, and must at one time have been deemed nearly inaccessible, it had all the appearance at the time of my survey of having remained in the same state for nearly two centuries. The quantity of slags I computed at from 8000 to 10,000 tons. If it is assumed that this furnace made upon an average annually 200 tons of pig iron, and that the quantity of slag run from the furnace was equal to one-half the quantity of iron made, we shall have 100 tons of cinders annually, for a period of from 80 to 100 years. If the abandonment of this furnace took place about the year 1640, the commencement of its smeltings must be assigned to a period between the years 1540 and 1560."

The oldest piece of cast iron which Mr. Mushet states he ever saw, exhibited the arms of England, with the initials E. R., and bore date 1555 (?), but he found no specimen in the Forest earlier than 1620. A few cast-iron fire-backs have been noticed in some of the old houses in the vicinity of the Forest, but none have an earlier date on them. The cast-iron grave-slabs found in the ancient iron-making districts of Surrey and Sussex do not occur here. He also observes that "although he had carefully examined every spot and relic in Dean Forest likely to denote the site of Dud Dudley's enterprising but unfortunate experiment of making pig-iron with pit-coal," no remains had been found. It was the same with the like operations of Cromwell, Major Wildman, Captain Birch, and other of his officers, doctors of physic and merchants, by whom works and furnaces had been set up in the Forest at a vast charge.

The troubles of the civil wars, in which the country surrounding the Forest was so much involved, materially disturbed its iron manufactures. Sir John Winter's large works at Lydney were wholly de-destroyed, and probably such others as continued in operation were limited to the casting of cannon and shot, similar to what was used in the siege of Goodrich Castle by Colonel Birch in 1646. Otherwise iron making was for the time suppressed.

When matters had become somewhat settled, the attention of the Commonwealth was directed to them. They were placed under the general supervision of Major John Wade, who was assisted in their management by John a Deane.

A document exists giving a debtor and creditor account from 13th September, 1653, to 20th August, 1655.[1] During these two years, upwards of £12,607 16s. 9¾d. was laid out by the Council of State and the Commissariat of the Admiralty, whilst only £10,705 14s. 3d. was received, leaving a deficit of £1902 2s. 6¾d.

Another paper states "what iron in pigs, barr, and shott have beene cast and made, sold, or otherwise disposed of, or remaining in stock," between 28th February, 1653, and 2nd August, 1656.

There remains also "a true inventory of all the tooles and utensils belonging to the forge at Whitecroft, this 13th August, 1656," divided into "all the chaffery, for the upper finery, for the lower finery."

John a Deane died in 1655, and was succeeded by Mr. John Roades.[2] From 2nd August, 1656, to 15th September, 1657, the Government account stood thus

	£	s.	d.
Dr. side	10,135	15	10¾
Cr.	8,023	15	3¼
Balance	2,112	0	7½

Hardly had the king's return been effected when, amidst the innumerable petitions which instantly greeted him, is one from Sir Hugh Middleton, Bart., for "the place of Overseer and Receiver of Profits of His Majestie's Iron Works in the Forest of Dean."[3] He strengthened his application with the timely remark that the appointment for which

---

[1] State Paper Office, Domestic Series, No. 835, fos. 675-710.
[2] Ibid., Domestic Series, Int., No. 816.
[3] State Papers, Domestic Series.

he sought was held by Major John Wade, "put in by Cromwell; an officer of which Wade, in July last (1659), robbed him of horses, arms, &c., kept him four months in close imprisonment for adhering to His Ma^{ty}, & has several times ransacked his house."

A contemporary petition, to much the same end, but from a different quarter, was presented by Sir Edward Massey. He stated, truly enough, that "he had formerly held the works for which he now applied, but they and all his stock were taken from him by the Rump Par-Parliament for his loyalty." But he suppressed saying, how they were formerly voted to him by the House of Commons for defeating the staunch royalist Sir John Winter, to whom they previously belonged.

Sir John, himself, was a third, and reasonable applicant for the restoration of his patent for the same, which was as justly restored him; the other, but unsuccessful candidate, being Sir Baynham Throckmorton.

In an elaborate return, [1] addressed to the Barons of the Exchequer, and dated the 12th April, 1662, the question is mooted, "What advantages will yearly accrue to His Ma^{ty} by his furnace and forge, if taken into his owne hands?" The answer is worked out in the following manner:—

"Imprimis.—Fower Long Coards of Wood will make two Loads of Coles wch two Loads of Coales will make one Tunne of Sowe Iron.

"Charges to make a Tunn of Sow Iron.

	li.	s.	d.
For cutting and coarding of Four Long-Coards genrlly will cost	00	14	00
For Coaleing at 3s 6d per loade	00	07	00
For carrying it to the Furnace, genrlly	00	07	00
For Mine and Oar	00	05	00
For Cinders	00	03	00
To the Founder or Caster	00	02	06
	01	18	06

"Price of a Tun of Sow Iron.

---

[1] Brit. Mus., Harl. MS. 6839, fol. 332.

"Which Tune of Sow Iron will yield cõib annis, although now debased by the late mispending of the Stock, but wil bee brought up ag^n to 6^{li} 10.

"What Quantity the Furnace will cast yearly.

"The Furnace may wth the Expence of 100^{li} to p^r serve & pcure a greater ppcõn of water, cast Were Thirty Tunn p weeke, but to reduce it to a greater certainty we will compute at 26 Tunne p weeke, wch at 6^{li} 10^s 6^d p Tunn amounts to 1248 Tunn p An., wch at 6^{li} 10^s 0^d p Tunn amounts to 8112^{li}. But the Charges to be deducted at 1^{li} 18^s 6^d p Tunn amounts to 2402^{li} 8^s deducting wch out of the generall pfitt there remains 709^{li} 12^s 0^d.

"Other Charges to be deducted and alowed out of the Furnace profits.

	li.	s.	d.
To a Stock Taker p An.	16	00	00
To a Clerk	40	00	00
To a Carpinter	6	13	04
Other Rep^{rs} of the Furnace (cõib annis)	12	00	00
For travelling Charges to the Clerk to sell iron	05	00	00
To two Wood Clerks p An.	20	00	00
And for Sacks and Hurdles p An.	20	00	00
Totall	119	13	04

"Charges of Product of the Forge.

"The Forge will make Cõibus Annis 150 Tunn wch will yield genrally 16^{li} 10^s p Tunn, although now debased by the late mispending of the Stock.

"Charges to make a Tunn of Barr Iron.

"Three Load of Coales will make a Tunn of Barr Iron, whereof one may be brasses, but sett it at three Loade,

	li.	s.	d.
The Cutting, Cording, Coaling, and Carriage will amount unto	02	02	00
And 2650 weight of Sow Iron will make one Tunn of Barr Iron, wch said 2650 weight of Sow Iron at 6^{li} 10^s p Tun amounts unto	08	12	00
And to the Workmen (viz.) Raffiners and Hammermen	01	00	00

Iron Making in the Olden Times

	11	14	00
Produce 150 Tunn at 16^{li} 10^s 0^d p Tun amounts p an. to	2475	0	0
Charges at 11^{li} 14^s 0^d p Tun as aforesaid amounts to	1740	0	0
Remaynes cleare	735	0	0

"Other Charges to be allowed out of the yearly pfitts of the Forge.

	li.	*s.*	*d.*
To a Clerk p Ann.	25	0	0
To a Stock taker	16	0	0
To a Carpinter	09	13	4
For other Rep^{rs}, as Oyle, Greese, &c., Cõibus Annis	20	0	0
	64	13	4

	li.	*s.*	*d.*
Totall of the Furnace, deducting the Officers' Fees, &c., is	5609	18	8
Totall of the Forge, deducting the Officers' Fees, &c., p An. is	0667	06	8
	6277	5	4

So considerable a balance each year, from one furnace and a single forge, admits of comparison with the profits made by ironmasters now.

The Commissioners further report that all necessary appliances existed on the spot:—

"One excellent Furnace called the Park Furnace, and one Forge called Whitecros Forge. The later is in good rep^{re}, but the Furnace wants a Roofe to ye Cole hous, and some other Rep^{rs}, wch we compute may cost us circa 40^{li}, and care must be taken whensoever his Ma^{ty} shall take them into his own handes, that all the Implem^{ts} the late psõns intrusted wth the managem^t thereof had deliv'd to them by inventory or otherwise, be forthcoming, or else it will be a great p^{r}judice to his Ma^{ty}."

It was also pointed out that, besides "the greate yearely pfitt" likely to accrue to the King, should he take the Iron Works into his own hands, they were "capable to serve his Navey both wth beter Iron and at much Easier Rates then now he payes for all sorts, and wee conceive that Iron Ordinance might be cast here for ye Service of ye Navey also at ye same rates." Some of the Forest iron, in the form of iron hoops, had already found its way to the navy store at Woolwich. [1]

Even the last winter's great storm (18th of February, 1662) is made to support their counsels, for the Commissioners affirmed that—"500^{li}, together with the young beechen timber lately blowne downe in the Lea Bayley, will sett the workes a goeing."

Lastly, the same officials suggested that a check should be put to the practice of sending iron ore and cinder out of the Forest, lest the supply to the king's works, as proposed, should run short. They suggest a tax "6d. at first, for fifteen bushells," adding "that they were informed that there is carryed out yearly at least 4000 dozen; and there is now lying at Newnham a small vessell to transport some for Ireland. There must needs be a Prohibition to carry out of the Forrest any cinders, least his Ma$^{ty's}$ owne works should need them in tyme." [2]

Reasons so carefully analyzed for inducing the Crown to take in hand iron making at Park End, deserved a better fate. But the king had irons enough in the fire, without becoming a manufacturer of iron in the Forest of Dean. Its timber was rather wanted for the navy, which the Duke of York longed to render more effective. Besides, places more convenient of access, in Surrey and Sussex, were supplying the iron trade. Hence, when in 1683 the above-named proposal was renewed by Sir John Erule, the Forest supervisor, it was rejected, although he promised a profit of £5390 per annum. [3]

The authorities went further than this, in refusing, as they thought, to sacrifice the timber for the iron. They even directed, about this time, the demolition of the Forest furnaces, thus reducing its iron works to such a degree as almost to annihilate them for the next hundred years.

---

[1] State Papers, Domestic Series.
[2] On similar principles, the eighth Order of the Free Miners' Court enacted that "no iron ore intended for Ireland should be shipped on the Severn or Wye for a less sum than 6s. 6d. for every dozen bushels."
[3] Commissioners' Report of 1788.

Iron Making in the Olden Times

What their recent state of prosperity had been, Andrew Yarranton, in his book of novel suggestions for the "Improvement of England by Sea and Land," printed in 1677, describes as follows:—

"And first, I will begin in Monmouthshire, and go through the Forest of Dean, and there take notice what infinite quantities of raw iron is there made, with bar iron and wire; and consider the infinite number of men, horses, and carriages which are to supply these works, and also digging of ironstone, providing of cinders, carrying to the works, making it into sows and bars, cutting of wood and converting into charcoal. Consider also, in all these parts, the woods are not worth the cutting and bringing home by the owner to burn in their houses; and it is because in all these places there are pit coal very cheap. . . . If these advantages were not there, it would be little less than a howling wilderness. I believe if this comes to the hands of Sir Baynom Frogmorton and Sir Duncomb Colchester, they will be on my side. Moreover, there is yet a most great benefit to the kingdom in general by the sow iron made of the ironstone and Roman cinders in the Forest of Dean, for that metal is of a most gentle, pliable, soft nature, easily and quickly to be wrought into manufacture, over what any other iron is, and it is the best in the known world; and the greatest part of this sow iron is sent up Severne to the forges into Worcester, Shropshire, Staffordshire, Warwickshire and Cheshire, and there it's made into bar iron: and because of its kind and gentle nature to work, it is now at Sturbridge, Dudley, Wolverhampton, Sedgley, Wasall and Burmingham, and there bent, wrought, and manufactured into all small commodities, and diffused all England over, and thereby a great trade made of it; and, when manufactured, into most parts of the world. And I can very easily make it appear, that in the Forest of Dean and thereabouts, and about the material that comes from thence, there are employed and have their subsistence therefrom no less than 60,000 persons. And certainly, if this be true, then it is certain it is better these iron works were up and in being than that there were none. And it were well if there were an Act of Parliament for enclosing all common fit or any way likely to bear wood in the Forest of Dean and six miles round the Forest; and that great quantities of timber might by the same law be there preserved, for to supply in future ages timber for shipping and building. And I dare say the Forest of Dean is, as to the iron, to be compared to the sheep's back as to the woollen; nothing being of more advantage to England than these two are. . . .

"In the Forest of Dean and thereabouts, the iron is made at this day of cinders, being the rough and offal thrown by in the Romans' time; they then having only foot blasts to melt the iron stone; but now, by the force of a great wheel that drives a pair of bellows twenty feet long, all that iron is extracted out of the cinders, which could not be forced from it by the Roman foot blast. And in the Forest of Dean and thereabouts, and as high as Worcester, there are great and infinite quantities of these cinders, some in vast mounts above ground, some underground, which will supply the iron works some hundreds of years, and these cinders are they which make the prime and best iron, and with much less charcoal than doth the ironstone. . . . Let there be one ton of this bar-iron made of Forest iron, and £20 will be given for it."

The 4th "Order" of the Mine Law Court, dated 27th April, 1680, fixes the prices at which twelve Winchester bushels of iron mine should be delivered at the following places:—St. Wonnarth's furnace, 10s.; Whitechurch, 7s.; Linton, 9s.; Bishopswood, 9s.; Longhope, 9s.; Flaxley, 8s.; Gunnsmills (if rebuilt), 7s.; Blakeney, 6s.; Lydney, 6s.; at those within the Forest (if rebuilt), the same as in 1668; Redbrooke, 4s. 6d.; the Abbey (Tintern), 9s.; Brochweare, 6s. 6d.; Redbrooke Passage, 5s. 6d.; Gunnpill, 7s.; or ore (intended for inland) shipped on the Severn, 6s. 6d.

Most of these localities present traces of long continued iron manufacture, especially St. Wonnarth's, Whitchurch, Bishopswood, and Flaxley, where the energetic proprietress, Mrs. Boevey, is said by Sir R. Atkyns to have had (c. A.D. 1712) "a furnace for casting of iron, and three forges." Charcoal is the only fuel of which any indications remain, the coppice woods being in several instances preserved from which it used to be obtained, and the furnaces are shown to have been invariably situated where waterpower was at command.

The prices affixed to the ore, including delivery, indicate a discontinuance, in a measure, of the mines on the north-east edge of the Forest. Those adjoining Newland and in Noxon Park, both on the op-

Iron Making in the Olden Times

posite side of the Forest, appear to have formed the principal sources of supply. The records of the Court of Mine Law, belonging to this date, allude oftener to these works than to others, for the same reason.

Its "order," dated 8th December, 1685, in providing that "the one-half of the jury of 48 should be iron-miners, and the other half colliers," manifests considerable decay in the influence and number of the former operatives, once so much otherwise. It is remarkable that the later orders are silent as regards iron, owing to the suppression of the Forest furnaces.

With respect to the mode now in use of reducing the mine ore, there is preserved so explicit an account, from the pen of Dr. Parsons, the county antiquary and naturalist of that age, as to call for its verbatim insertion here:—

> "The ore and cinder, wherewith they make their iron (which is the great employment of the poorer sort of inhabitants), 'tis dug in most parts of the Forest, one in the bowells, and the other towards the surface of the earth.
>
> "There are two sorts of ore: the best ore is your Brush ore, of blewish colour, very ponderous, and full of shiny specks, like grains of silver; this affordeth the greatest quantity of iron, but being melted alone, produceth a metal very short and brittle. To remedy this inconvenience, they make use of another material, which they call cinder, it being nothing else but the refuse of the ore, after the melting hath been extracted, which, being melted with the other in due quantity, gives it that excellent temper of toughness for which this iron is preferred before any other that is brought from foreign parts.
>
> "After they have provided their ore, their first work is to calcine it, which is done in kilns, much after the fashion of our ordinary lime kilns; these they fill up to the top with coal and ore untill it be full, and so, putting fire to the bottom, they let it burn till the coal be wasted, and then renew the kilnes with fresh ore and coal. This is done without any infusion of mettal, and serves to consume the more drossy part of the ore, and to make it fryable, supplying the beating and washing, which are to no other mettals; from hence they carry it to their furnaces, which are built of brick and stone, about 24 foot square on the outside, and near 30 foot in hight within, and not above 8 or 10 foot over where it is widest, which

is about the middle, the top and bottom having a narrow compass, much like the form of an egg. Behind the furnace are placed two high pair of bellows, whose noses meet at a little hole near the bottom: these are compressed together by certain buttons placed on the axis of a very large wheel, which is turned round by water, in the manner of an over-shot mill. As soon as these buttons are slid off, the bellows are raised again by a counterpoise of weights, whereby they are made to play alternately, the one giving its blast whilst the other is rising.

"At first they fill these furnaces with ore and cinder intermixt with fuel, which in these works is always charcoal, laying them hollow at the bottom, that they may the more easily take fire; but after they are once kindled, the materials run together into an hard cake or lump, which is sustained by the furnace, and through this the mettal as it runs trickles down the receivers, which are placed at the bottom, where there is a passage open, by which they take away the scum and dross, and let out their mettal as they see occasion.

"Before the mouth of the furnace lyeth a great bed of sand, where they make furrows of the fashion they desire to cast their iron: into these, when the receivers are full, they let in their mettal, which is made so very fluid by the violence of the fire that it not only runs to a considerable distance, but stands afterwards boiling a great while.

"After these furnaces are once at work, they keep them constantly employed for many months together, never suffering the fire to slacken night or day, but still supplying the waste of fuel and other materials with fresh, poured in at the top.

"Several attempts have been made to bring in the use of the sea coal in these works instead of charcoal; the former being to be had at an easy rate, the latter not without a great expence; but hitherto they have proved ineffectual, the workmen finding by experience that a seacoal fire, how vehement soever, will not penetrate the most fixed parts of the ore, by which means they leave much of the mettal behind them unmelted.

"From these furnaces they bring the sows and piggs of iron, as they call them, to their forges; these are two sorts,

Iron Making in the Olden Times

though they stood together under the same roof; one they call their finery, and the other chafers: both of them are upon hearths, upon which they place great heaps of sea coal, and behind them bellows like those of the furnaces, but nothing near so large.

"In such finerys they first put their piggs of iron, placing three or four of them together behind the fire, with a little of one end thrust into it, where softening by degrees they stir and work them with long barrs of iron till the mettal runs together in a round masse or lump, which they call an half bloome: this they take out, and giving it a few strokes with their sledges, they carry it to a great weighty hammer, raised likewise by the motion of a water wheel, where, applying it dexterously to the blows, they presently beat it into a thick short square; this they put into the finery again, and heating it red hot, they work it under the same hammer till it comes to the shape of a bar in the middle, with two square knobs in the ends; last of all they give it other heatings in the chaffers, and more workings under the hammer, till they have brought their iron into barrs of several shapes, in which fashion they expose them to sale.

"All their principal iron undergoes the aforementioned preparations, yet for several other purposes, as for backs of chimneys, hearths of ovens, and the like, they have a sort of cast iron which they take out of the receivers of the furnace, so soon as it is melted, in great ladles, and pour it into the moulds of fine sand in like manner as they do cast brass and softer mettals; but this sort of iron is so very brittle, that, being heated with one blow of the hammer, it breaks all to pieces."

As an instance of the considerable extent to which the old cinders continued to be used in the iron furnaces round the Forest, the following abstract of an indenture, found in Mr. Wyrrall's collection, and dated 20th October, 1692, may be quoted:—

"Jephthah Wyrall, Gent., to R^d Avenant, Gent., and John Wheeler, Gent.

"Articles for the Sale of 10 thousand doz^n of cinders, in certain grounds near Mr. Wyrall's house, called the Correggio, the Limekiln Patch, the Long Sevens, and the

Ockwal Field, if so many could be found there. The Price, 10 Pence the dozen, or 12 Bushels; 6 to be heaped and the other 6 even with the top of the Bushel, or hand-weaved. Such of them as should be taken to Bishopswood or Parkend to be measured by the Bushel used at Bishop's wood Furnace; and such as should be carried to Blakeney Furnace by the Bushel used there. To be raised and fitted for carriage by Avenant and Whealer. To employ no persons in raising the cinders but such as Mr. Wyrall approves of. Mr. Wyrall to carry yearly as many cinders as he should please, not exceeding 250 Dozens, to Parkend, at 4^s a dozen. Should carry to the banks of the river Wye, at 13^d a Dozen such as should be used at Bishop's Wood Furnace. Avenant and Whealer to get 800 doz^n a year, and as many more as they sh^d please till the 10 Thousand Dozens should be raised: and pay for them yearly on the 1st day of May, and the 1st day of October; and should leave the ground as level and plain as usually is where cinders are gotten (which was promising nothing at all)."

According to a paper examined by Mr. Mushet, and referring probably to the year 1720 or 1730, the iron-making district of the Forest of Dean contained ten blast furnaces, viz., six in Gloucestershire, three in Herefordshire, and one at Tintern, making their total number just equal to that of the then iron-making district of Sussex. In Mr. Taylor's map of Gloucestershire, published in 1777, iron furnaces, forges, or engines are indicated at Bishopswood, Lydbrook, the New Wear, Upper Red Brook, Park End, Bradley, and Flaxley. Yet only a small portion of the mineral used at these works was obtained from the Dean Forest mines, if we may judge from the statement made by Mr. Hopkinson, in 1788, before the Parliamentary Commissioners, to the effect that "there is no regular iron mine work now carried on in the said Forest, but there were about twenty-two poor men who, at times when they had no other work to do, employed themselves in searching for and getting iron mine or ore in the old holes and pits in the said Forest, which have been worked out many years." Such a practice is well remembered by the aged miners, the chief part of the ore used in the above-named furnaces having been brought by sea from Whitehaven.

Iron Making in the Olden Times

[1] Thus Mr. Mushet represents, "at Tintern the furnace charge for forge pig iron was generally composed of a mixture of seven-eighths of Lancashire iron ore and one-eighth part of a lean calcareous sparry iron ore, from the Forest of Dean, called flux, the average yield of which mixture was fifty per cent. of iron. When in full work, Tintern Abbey charcoal furnace made weekly from twenty-eight to thirty tons of charcoal forge pig iron, and consumed forty dozen sacks of charcoal; so that sixteen sacks of charcoal were consumed in making one ton of pigs." This furnace was, he believes, "the first charcoal furnace which in this country was blown with air compressed in iron cylinders."

Flaxley was one of the very last places where iron was made in the old way. The Rev. T. Budge, writing at the commencement of the present century, says of it:—

> "The iron manufactory is still carried on, and the metal is esteemed peculiarly good; but its goodness does not arise from any extraordinary qualities in the ore, but from the practice of working the furnace and forges with charcoal wood, without any mixture of pit-coal.
>
> "The quantity of charcoal required is so considerable that the furnace cannot be kept in blow or working more than nine months successively, the wheels which work the bellows and hammers being turned by a powerful stream of water. At this time (28th Oct. 1802) a cessation has taken place for nearly a year. Lancashire ore, which is brought to Newnham by sea, furnishes the principal supply; the mine found in the Forest being either too scanty to answer the expense of raising it, or when raised too difficult of fusion, and consequently too consumptive of fuel, to allow the common use of it.
>
> "When the furnace is at work, about twenty tons a week are reduced to pig iron; in this state it is carried to the forges, where about eight tons a week are hammered out into bars, ploughshares, &c., ready for the smith."

---

[1] To these works Mr Thoresby alludes, in his diary, 7 Sept., 1694, recording that near Egremont he passed "by the iron mines, where we saw them working, and got some ore."

Though these operations have been long given up, the furnace buildings removed, and the pools drained in which the water accumulated for driving the machinery, yet the old people of the neighbourhood still recollect when the Castiard's Vale, now wholly devoted to the picturesque, resounded with the noise of engines. A solitary heap of Lancashire iron mine alone remains to show what was once operated upon at this spot.

The year 1795 marks the period when the manufacture of iron was resumed in the Forest by means of pit-coal cokes at Cinderford, the above date being preserved on an inscription stone in No. 1 furnace. "The conductors of the work succeeded," in the words of the late Mr. Bishop, communicated to the author,—

> "As to fact, and made pig iron of good quality; but from the rude and insufficient character of their arrangements, they failed commercially as a speculation, the quantity produced not reaching twenty tons per week. The cokes were brought from Broadmoor in boats, by a small canal, the embankment of which may be seen at the present day. The ore was carried down to the furnaces on mules' backs, from Edge Hill and other mines. The rising tide of iron manufacture in Wales and Staffordshire could not fail to swamp such ineffectual arrangements, and as a natural consequence Cinderford sank.
>
> "Attempts still continued to be made from time to time in the locality, but the want of success, and the loss of large capital, placed the whole neighbourhood under a ban.
>
> "Moses Teague was the day-star who ushered in a bright morning after a dark and gloomy night. Great natural genius, combined with a rare devotion to the interests of the Forest, led him to attempt a solution of the difficulty. In this he so far succeeded that he formed a company, consisting of Messrs. Whitehouse, James, and Montague, who took a lease of Park End Furnace about the year 1825, erected a large water-wheel to blow the furnace, and got to work in 1826. Having started this concern, Mr. Teague, who from constitutional tendencies was always seeking something new, and considered nothing done while aught remained to do, cast his eye on Cinderford, which he thought presented the best prospects in the lo-

cality; and after making arrangements with Messrs. Montague, Church, and Fraser, those gentlemen with himself formed the first 'Cinderford Iron Company,' the writer joining the undertaking when the foundations of the buildings were being laid. The scheme comprehended two blast furnaces, a powerful blast engine still at work, finery, forge, and rolling-mill, designed to furnish about forty tons of tin-plate per week, with collieries and mine work. Before the completion of the undertaking it was found that the outlay so far exceeded their expectations and means that the concern became embarrassed almost before it was finished, which, with the then great depression of the iron trade during the years 1829 to 1832 inclusive, led to the stoppage of the works, which had continued in operation from November, 1829, till the close of 1832, in which state they continued to 1835, when Mr. Teague again came to the rescue, and induced Mr. William Allaway, a gentleman in the tin-plate trade, of Lydbrook, to form, in connexion with Messrs. Crawshay, another company. Mr. Teague having retired from the management of the furnaces, that important post was filled by Mr. James Broad, a man of great practical knowledge, who for twenty years succeeded in making iron at Cinderford Furnaces of quality and in quantities which had never been anticipated. There are now four blast furnaces, three of which are always in use, and a new blast engine of considerable power is in course of erection, in addition to the old engine, which has been puffing away for twenty-eight years."

As narrated in an earlier part of this account, Park End long since possessed a furnace and forge, though afterwards suppressed in 1674, and not resumed until 1799, the date of the oldest iron furnace there. It is situated about half a mile lower down the valley than the former one, and was carried on by a Mr. Perkins. The Works were eventually sold to Mr. John Protheroe, and by him disposed of to his nephew, Edward Protheroe, Esq., formerly M.P. for Bristol, who was likewise the possessor of several collieries near. In 1824 Mr. Protheroe granted a lease of the furnace and premises, and also sundry iron mines, to the Forest of Dean Iron Company, then consisting of Messrs. Montague, James, & Co. This arrangement continued until 1826, when Messrs. William Montague, of Gloucester, and John

James, Esq., of Lydney, became the sole lessees. A second furnace was erected by these gentlemen in 1827, as well as an immense waterwheel of 51 feet diameter and 6 feet wide, said at the time to be the largest in the kingdom. Two extensive ponds, still observable, were formed higher up the vale, and connected with the Works by a canal yet remaining. Little use was made, however, of these appliances, owing to the general introduction and superior advantages of steam power. A steam-engine was consequently put up for creating the necessary blast. Not being found sufficiently powerful to keep two furnaces in operation, each being 45 feet high, 9½ feet diameter at the top, 14 feet across at the boshes, and 5 feet diameter at the hearth, another steam-engine of 80 horse power was erected in 1849; but owing to a depression in the iron trade, and other causes, the two furnaces were not then worked together.

A few years after the decease of Mr. Montague, in 1847, Mr. James bought all his interest in the Works and became the sole lessee, until the year 1854, when he purchased from Mr. Protheroe the fee of the property, together with all the liabilities of the lease. Since that time the two furnaces have been occasionally worked together, under the superintendence of Mr. Greenham, one of the proprietors, the firm still continuing as "The Forest of Dean Iron Company." They produce upwards of 300 tons of pig-iron per week, consuming in the meantime 350 tons of coke, and 600 tons of iron ore, obtained from the neighbouring mines at Oakwood and China Eugene; and from the Perseverance and Findall Mine, on the eastern side of the Forest. These operations give employment to something like 300 men; and the foundation is now being laid for another furnace.

Besides its iron furnaces, Park End is the site of Messrs. T. and W. Allaway's extensive Tin-Plate Works, erected at a large outlay by Messrs. James and Greenham in 1851. They find employment for some 200 work-people, by whom 500 boxes of tinplate are made per week. Two-thirds of the iron so used is obtained in the Forest.

Similar works, only on a larger scale, are carried on at Lydney by Messrs. W. Allaway and Sons. These are five in number, and bear the names of The Lower Mill, The Lower Forge, The Middle Forge, The Upper Mill, and The Upper Forge. About 400 hands are engaged at them, and turn out about 1,000 boxes of tin-plate every week, besides a quantity of sheet-iron. The materials supplied to these works from the Forest of Dean are pig-iron, coal, fire-bricks and clay, fire-stone and fire-sand, and cordwood for conversion into charcoal. Lydney has long been famed for its ironworks, which at one time belonged to the Talbot family.

Sowdley, in spite of its natural beauties and retired situation, has been occupied by ironworks since 1565, the ancient family of the Joneses of Hay Hill conducting them as wire-works drawn by power of hand. Messrs. Parnell and Co. then took to them; from 1784 to 1804 Dobbs and Taylor carried on the works; Browning, Heaven and Tayer followed in 1824, and Todd, Jeffries and Spirrin in 1828, converting a part of the premises into paint and brass works.

In 1837 they were raised to the dignity of blast furnaces by having two of them erected of the usual size, by Edward Protheroe, Esq., and worked by him for four years. The late Mr. Benjamin Gibbons purchased them in 1857; and in 1863 his representatives sold them to Messrs. Goold, by whom they are conducted. At present but one furnace is in blast, yielding about 20 tons of Forest iron each casting, South Wales coke being the fuel employed. Eighty hands are engaged at these works.

Lydbrook has long been the site of several busy ironworks. They may be specified as the Upper and Lower; the last of these, situated near the Wye, was once the property of the Foleys, by whom so many of the iron works of the beginning of the last century were carried on. More recently they were in Mr. Partridge's hands, and were worked in connexion with the furnace at Bishopswood. In 1817 Mr. Allaway leased them, at which time they comprised three forges, rolling and bar mills, and tin-house complete, capable of producing from 100 to 150 boxes of tin plates per week. Now, however, under the able management of the late Mr. Allaway's sons, the Works yield 600 boxes, sent off by the Wye. The iron used is chiefly that from Cinderford, as being the best suited for the purpose.

The Upper Works, formerly the property of Lord Gage, at the time when the High Meadow Estates belonged to the family, are now owned by Messrs. Russell, the late Mr. Russell having bought them from the Crown in 1818. His son, Mr. Edward Russell, writes:—

> "We have since then considerably improved and enlarged them, and are now employing about 100 hands. We manufacture wire for fencing, as also for telegraph purposes, of which we can roll from 40 to 50 tons per week. We likewise make charcoal iron for horse-nails and smith's work, besides that for agricultural purposes, using the Cinderford, Shropshire, and Staffordshire iron, especially the former."

Other works, resembling those just described, are being carried on by Mr. James Russell at the Forest Vale Iron Works, near Cinderford. When perfected, they will employ not less than 60 pairs of hands, and will supply considerable quantities of iron rods for telegraphic and other wire, as well as chain-cable iron, the adjoining furnaces affording the requisite metal.

All the iron ore supplied from this neighbourhood to these different works is derived from one or other of the following iron mines, whose present extent may be thus particularized. [1]

The *Shake-mantle, Buckshraft,* and *St. Annal's* pits, on the *eastern* side of the Forest, constitute that exceedingly important range of mining operations, from which the Cinderford furnaces have long obtained their chief supply of iron ore. These are four in number, having a height of 43 feet, an extreme breadth of 14 feet, that of the hearth being 6 feet. They make 500 tons every week of the finest hot-blast iron.

A peculiar interest attaches to the first of these three pits, owing to its being the oldest mine still at work in this vicinity, though it dates no earlier than 1829, so recently has iron mining been resumed in this part of the district. Buckshraft was begun in 1835-7, and that at St. Annal's in 1849, each originating in the increasing demand for iron ore at the adjoining blast furnaces. They all descend to the same vein of red hematite, as well as to one common "level." This runs from one to the others, almost in a direct line two miles long. The shafts are severally 70, 160, and 221 yards deep.

Upwards of 36,000 tons of rich ore have been annually obtained from these iron mines for many years, leaving a transverse area of cavernous workings about 70 yards wide. But a far greater void was formed by the old miners, whose holes occur immediately above, and in which a few scattered tools have been discovered, left behind when operations were abruptly stopped in 1674, but not before the men had burrowed down some 150 yards.

The natural drainage of these mines being towards the Shake-mantle pit, a very powerful pumping engine has been put up there, capable of raising 250 gallons of water to the surface at every stroke.

As many as 250 hands are employed in working these valuable iron mines.

---

[1] There are other important Iron works at Tintern, Redbrook, &c., but it does not appear that Dean Forest iron is used at them.

Iron Making in the Olden Times

The *Westbury-brook* iron mine, so called from its situation near the head of that stream, is one of the most productive pits on the *eastern* side of the Forest basin.

It was begun about the year 1837, immediately below "the old men's workings." These proved to be remarkably extensive and searching, all the ore having been cleared out to a depth, in some places, of 160 yards. They were also found to contain many ancient mining implements, such as plank-ladders, shovels, helves, &c., all of ash, besides leather shoes and mattock heads, left behind probably when the iron furnaces of the district were suppressed in 1674.

Since 1843 this mine work has been very prosperously conducted by the agents of the Dowlais Iron Company, whither most of its ore is sent to be mixed and smelted with the ore there, much to the improvement of the iron so made.

Nearly 200 hands are employed at the Westbury-brook mine pit. The excavations run north and south for upwards of a mile and a half, their breadth averaging about 16 yards. They are reached by a shaft 186 yards deep, to the top of which a plunging pump raises 33 gallons of water at each stroke.

For several years past this iron mine has yielded many thousands of tons yearly of the finest red hematite ore. A steam-engine of 36 horse power brings it to the surface.

The *Old Sling* iron mine, begun in 1838, on the Clearwell Mean, has long been considered one of the principal mine works on the western edge of the Forest. Its chief access is by a shaft that descends 105 yards to where the deepest workings begin. These gradually rise, in accordance with the upward slope of the mine train, until they attain an area of about 20 acres, leaving some 33 acres unwrought above them, to where "the old men's workings" are reached. Such is the case about 50 yards below the surface, after they had worked over upwards of seven acres of the mine ore. These excavations were found to contain some ancient picks and wooden shovels tipped with iron, an addition not met with elsewhere, but rendered necessary in this instance by the harder nature of the matrix of the mine ore.

This iron mine has yielded for several years past 1,000 tons of red hematite ore per month, and employed nearly 100 hands.

Another remunerative iron mine, opened on the western side of the forest, is the *Easter* iron mine. It has three shafts sunk upon it, 100, 113, and 118 yards deep respectively. The first of these, and the only one in work, at which a light steam-engine of 14 horse power is used, communicates with "the old men's workings," though none of their tools have been found in them. About fifty men and boys are employed in this mine, from which upwards of 1,000 tons of ore are procured each month.

The table here appended, by the kind permission of the deputy gaveller, Mr. T. F. Brown, exhibits the proceeds of each of the Dean Forest Iron Mines during the years 1864-5:—

AN ACCOUNT OF IRON ORE RAISED IN DEAN FOREST AND HUNDRED OF ST. BRIAVEL'S FROM CHRISTMAS, 1863, TO CHRISTMAS, 1865.

NAME OF IRON MINE.	Half-year ended Mid Summer 1864.	Half-year ended Christmas 1864.	Total.	Half-year ended Mid Summer 1865.	Half-year ended Christmas 1865.	Total.
Perseverance and Findall	5,199	4,217	9,416	5,742	7,126	12,868
New China Level	123	66	189	240	170	410
New Dun Pit	1,255	985	2,190	...	...	...
Buckshraft	21,400	18,370	39,770	22,245	23,882	46,127
Tingle's Mine Level	548	...	548	...	405	405
Crow's Nest	1,893	2,975	4,868	...	...	...
Old Ham	514	...	514	89	456	545
Oakwood Mill	2,923	2,222	5,145	1,723	4,761	6,484
Westbury Brook	10,180	9,773	19,953	7,756	11,293	19,049
Old Sling	8,889	7,051	15,940	6,267	6,113	12,380
Easter	5,584	3,911	9,495	1,788	2,760	4,548
Yewtree	173	67	240	...	...	...
Dean's Meend	7,540	7,228	14,768	8,192	6,176	14,368
Clearwell	1,277	3,416	4,693	...	...	...
Shraves	731	364	1,095	367	186	558
Scar Pit	524	...	524	...	...	...
Staunton	...	...	...	543	941	1,484
Wigpool	...	...	...	...	402	402
Scar Pit	...	488	488	...	...	...

Iron Making in the Olden Times

Forty other gales of iron ore have been awarded to various parties, and will no doubt be shortly opened.

No account of the production of iron in the Forest of Dean can be called complete which does not include some description of the "laws and privileges," the "customs and franchises" of the original operatives by whom the mine ore was obtained. As the miners themselves invariably refer to the "Book of Dennis" and the seventeen orders of their court of mine law for all authoritative information respecting their guild, or fraternity of free minership, the reader is furnished with the following summary of their contents.

Thus the first-named document begins by specifying the franchises of the mine locally and personally, meaning its liberties or privileges, as not to be trespassed against, and consisting apparently in this, that every man who possessed it, *though it is not stated how*, might, with the approval of the king's gaveller, dig for iron ore or coal where he pleased, not limiting him, as in later times, to the Hundred of St. Briavel's, but giving as his range the whole county south-west of Gloucester and as far south as the Severn. There was, too, a right of way awarded to every mine, although in certain cases "forbids" to sell might be declared.

One-third part of the profits of the undertaking belonged to the king, whose gaveller called at the works every Tuesday "between Mattens and Masse," and received one penny from each miner, the fellowship supplying the Crown with twelve charges of ore per week at twelve pence, or three charges of "sea coal" at one penny.

Timber was allowed for the use of the works above and below ground.

Only such persons as had been born and were abiding in the Forest were to frequent the mines, in working which the distance of a stone's throw was always to be kept, and property in them might be bequeathed.

The miners' clothes and light are mentioned, as likewise the standard measure called "bellis," and carts and waynes are prohibited.

It alludes to the "court of the wod" at the speech before the Verderers; but more particularly to the debtor court at St. Briavel's castle or gate, and to the mine court, as regulated by the constable, clerk, and gaveller, with the miners' jury of twelve, twenty-four, or forty-eight, where all causes relating to the mines were to be alone heard. Three hands, or three witnesses, were required in evidence, and the oath was taken with a stick of holly held in the hand.

Iron Making in the Olden Times

The miners of Mitcheldeane, Little Deane, and Ruer Deane are called "beneath the wood." It also appears that at Carlion, Newport, Barkley, Monmouth, and Tulluh, the manufacture of iron was carried on by "smiths," who were connected with smith holders living in the Forest, and supplying the ore.

For many ages the mining operations of the Forest and the action of the miners' court seem to have gone on so smoothly, and as a matter of course, that no notices regarding them occur in the documents of those times.

With the Restoration, however, and the revival of the ancient rights of the crown, it was found necessary to resume the sessions of the court of mine law, under the presidency of Sir Baynham Throckmorton. Thus it first of all met again on the 16th November, 1663, and continued so to do, from time to time, for the ensuing Hundred years, passing at different periods its seventeen "orders." These verdicts are chiefly remarkable for reducing the area of the miners rights to the Hundred of St. Briavel's, though they fail to say what constituted *free minership* beyond the old definition given in the "Book of Dennis," viz., "beene borne and abiding within the castle of St. Brevill's and the bounds of the Forest as aforesaid." In 1834 the Government commissioners were informed that it involved birth from a free father, and working a year and a day in the mines. They are still a numerous and important fraternity, without whom no new mine works can be commenced.

Effigy of a Forest Free Miner, reduced from a Brass of the 15th century in Newland Church.

Their aspect when accoutered for work is given in the frontispiece. If compared with their mediæval appearance, as displayed in the miners' crest, the interval of four hundred years is scarcely discoverable. Every mining appurtenance is retained, only somewhat altered in shape, and that, perhaps, not for the better, be it cap, "bellis," or general attire. Only the beard is absent, but then there are the shoes.

Iron Making in the Olden Times

Forest of Dean Iron Miners ready for work, from a Photograph by the late Mr. Atkinson.

On several occasions they conferred their freedom on the leading gentry of the neighbourhood. By their orders they also sanctioned the sinking of *pits*, as distinguished from *levels*, extending the interval between mine and mine from "within so much space that ye miner may stand and cast ridding and stones soe farr from him with a bale as the manner is," to five hundred yards. At the present time the deputy gaveller, Mr. T. Forster Brown, is the resident official under the Commissioner in charge of Her Majesty's Woods, &c., and he, with his respected predecessor, have at all times most obligingly facilitated the author's inquiries by giving the desired information. It was during the deputy gavellership of the late Mr. John Atkinson at Coleford that the writer chanced to meet with the original transcript, here presented to the reader, of the "Book of Dennis." The first printing and publication of it took place in 1687, by William Cooper, at the Pelican, in Little Britain, and it has been frequently but imperfectly reprinted.

Finding on examination that the reign of the first of the Edwards, and not the third, was the period to which it assigned the confirmation of the Forest of Dean Mine Laws, and that it contained many other inaccuracies, he determined to prepare, in accordance with the valued suggestion of Mr. Smirke, Judge of the Stannaries of Cornwall, a true copy of so ancient and curious a document.

From the note which is appended to it, the existing MS. is evidently the only authentic copy of the original "parchment roll," out of which it was transcribed by the gaveller, Richard Morse, A.D. 1673, of the penmanship of which period it is a good specimen.

It seems to be a presentment of the Court of Mine Law, duly signed by the jury of forty-eight free miners. Although its early date, and one or two forms of expression, may seem to indicate that it was first of all written in Latin, yet so many of its words and phrases, together with its concluding signatures, are so thoroughly old English, as to show that it was most probably composed in our own language. There are no paragraphs nor punctuations.

In character it is "sui generis," though it exhibits traces of resemblance to the laws and customs of the old mining districts of Somerset and Derbyshire, and even with those of Germany, as the accompanying notes show. The words between brackets do not occur in the original MS., having been inserted by modern printers. Those in italics give the corrections needed in modern copies.

THE MINERS LAWES AND PRIVILLEDGES.

Bee itt in minde and [in] Remembrance what the Customes and [the] Franchises *hath* been that were graunted tyme out of Minde and after in tyme of the Excellent and Redoubted Prince King Edward [1] [the Third] *un*to ye Miners of the Forrest of Deane and the Castle of St Bridvills and the bounds of the said**[Perambulations of the Mine.]** Forrest (That *is to say*) First [2] betweene Chepstowe Bridge and Gloucester Bridge the halfe *deale* of Newent Ross Ash Monmouth*s* bridge and soe farr in*to* the Seasoames as the Blast of a horne or the voice of a man may bee heard Soe that if any did Trespasse **[Miners' power to sue trespassers.]** against the Franchises of the Min*ers* [that is to say] that pass[ing] by boat [3] Trowe Pinard [4] or any other Vessell without gree [5] made for the Customes due to the King and also to the

---

[1] It is difficult to explain the bold introduction of so important an insertion, unless we attribute it to the over-wisdom of some modern printer, who regarded Edward III. as the only excellent and redoubted prince of the Edwardian category.

[2] These comprehensive limits mark an early age; but in mining matters they were hardly more than nominal—the mineral district comprising only a third of the land thus circumscribed.

[3] The proximity of the Severn, and particularly the Wye, to the mine works of the age is here shown.

[4] Printed "pichard," meaning, possibly, the Wye coracle.

[5] The French word "gree," for agreement or composition, is familiar among our early poets and writers, and occurs in the old statutes.

said Miners for the Myne [1] then hee that passe*th ought* [passed out] to come by the noyse of the horne or *the* cry And if hee will not come again Then his Boate or Vessell and all his Cattell **[Forfeiture.]** within forth beene forfeit *un*to the King for the Forbadment [2] broken the which is atta^{c}hme^t in the Franchises of the said Miners [and] Also [3] that the said **[Their power to mine in any place.]** Myners may myne in any place that they will as well without the bounds as within without *the* Forebodment of any man But if so [be] *that* a*ny* Smith [4] have a Smithman at Karleton [5] Newport or at Barkley th*en such* [which] Smithman is occupied [6] in Smith *craft* [work] and in Covenant with a Smith holder within the **[Covenant servant a fugitive.]** said Bounds Then the Smith holder [that is] within shall goe to the said Townes to prove his Covenant and after *his* [the] proffe *he may* [made] not have his Smithman Then ye Smith holder shall forbidd all the Myne that *ought* [might] to be carryed of the said Strainger that occupied the said Smithman unto the tyme that hee answereth as right is Then the [said] Smith holder within shall not forbidd the Myne of no other [man] but only of him that occupieth [occupied] the said Smithman Also in the said manner if any Smithman bee in Monmouth or Trellich then the Smith holder *within* shall come to St Briavell's Gate [7] and there with three hands [8] shall prove his Smithman and the prooffe made a precept shall bee delivered by the Constable to the Gaveller the which is Bayliffe of the said Myne to **[Gaveller is bayliffe of the mine.]** forbidd the Mine of him that occupieth the said

---

[1] In this and in several other passages of this document, "myne " is used for mineral or ore.
[2] This word and its variations is technical, and is nearly equivalent to a prohibition or injunction.
[3] This general liberty of mining, without apparent restriction as to surface ownership, is to be found in the earliest charters of the Stannaries, and was and still is extensively prevalent in Germany and elsewhere. The authorities are collected in Mr. Smirke's volume already referred to. It was this remarkable liberty that Lord Nelson noticed when visiting the Forest in 1802.
[4] In very early times the smith ranked very high among artificers, and was honoured in proportion.
[5] Probably carbon, old iron cinders, are still found at these places.
[6] The gate being the spot where justice was administered, in accordance with remote practice.
[7] Or Court of the Mine held in the castle.
[8] "Tertiâ manu," with a third hand; that is, with three witnesses or compurgators.

Smithman till hee bee restored and only of him and [of] noe other Also [1] the Miners have such libertyes and Franchises that for catelo [2] to them due for their Myne that they beene Bayliffs to take the Cattle of their **[Miners and bayliffe may arrest cattle for their debts.]** debtors and to arrest them without *the* leave of any man till gree bee made if hee bee within the bounds aforesaid And if the Debtor bee without the bounds in what place that hee bee Then the Miner shall forbidd all the Myne that ought to bee carryed to the place in wch the debtor bee abiding till Gree bee made to the Miner And after the forbodment if any carry [mine] to **[Forbode for debt due without the mine.]** the place aforesaid against the forbidd The Carrier shall be accountable and debt^r to the Miner as the principall was And alsoe the beasts that carry the Myne shall be forfeit to the King for the forbodd broken *And* [Also] if a Smith holder or any other bee debtor for *the* Myne *un*to a Myner the wch Smith holder or other bee within Then the Myner is Bayliff in every place (Except his own close) to take the horse of the *said* debtor if hee **[Distreyning a horse.]** bee saddled of a work saddle and of noe other saddle bee it that the horse bee halfe within the door of the Smith soe that the Myner may take the tayle of the horse The debtor shall deliver the horse to the Myner And [3] if hee [so] doe not the Myner shall [make and] levy *and* **[Hue and crye.]** *make* huy and cry ag^t the horse and then the horse shall bee forfeit to the King for the hue and cry made and levied And yet ye Miner shall present the debtor in the Mine Law *the* wch is Court for the Myne And the*re* the debtor before the Constable and his Clarke the Gaveller and the Miners and none other Folke to plead right *but* onely the Miner*s* shall bee there and hold a **[Holly sticks, &c.]** sticke of holly and then the said Myner demanding the debt shall putt his hand upon the [said] sticke and **[Swears his debt.]** none others with him and shall sweare *upon* [by] [4]

---

[1] In allusion to this rude and arbitrary process of distress, Mr. Smirke states that it is abundantly countenanced by ancient usage, especially in the Hartz Mines. Haltaus says—"Olim pignoris captio ex debitoris rebus moventibus diu privatorum arbitrio permissa."

[2] The "cattle" here must not be understood as exclusively applicable to live stock, it refers to all personal "chattels" or goods.

[3] However whimsical this claim may appear, observes Mr. Smirke, it is almost exactly paralleled in the law ascribed to Rob. I. of Scotland:—"Si debitor per vim a parte creditoris namos abstulerit, creditor cum sectâ vel huesis persequatur ablatorem."

[4] A copy of the Holy Gospels was eventually used on such occasions.

his Faith that the said debt is to him due and the prove made the debtor in the same place shall pay the Myner all the debt proved or els hee shall be brought to the Castle of St. Briavell's till gree bee made and also hee **[Amersement.]** shall be amersed to the king in two shillings and the same manner Myner to Myner and Myner to all other folke Also if a Distresse bee taken in like manner *as* aforesaid And the Debtor lett the distresse dye or bee impaired within ye Ward of the Myner for fraud or for malice and after the Myner shall distreyne and take **[Distresse.]** more distresse if any bee till Gree bee made And bee it that the distress dye or bee impaired within the ward of the Myne[r] the debtor shall not have right to implead the Miner neither noe right to grieve him for the Trespasse done But at all tymes the Myner ha*ve* [hath] right to take other distresse till gree be made Also for the Myne of an horse as is aforesayd the Miner **[Horse girth and halter.]** shall take the foregirth for three half-pence and for one penny the halter Also the Myner hath such franchises to enquire the Myne ^{1} in every soyle of the Kings of which it may be named ^{2} and also of all other Folke **[To dige in ye king's soyle or any other.]** without the with saying of any man and also if any bee that denyeth any soyle whatsoever hit bee bee hit sowed or noe or what degree hit may be named ^{3} Then the Gaveller by the strength of the King shall deliver the soyle to the Myners with a convenient way next **[Wayes to ye pitte.]** stretching to the King's highway by the wch Myne may be carried to all places and waters that been convenient to the sayd Myne without withsaying of any man ^{4} For the wch Soyle in [the] wch the myne is within found The Lord of the Soyle at the first time if hee will enter **[The lord of ye soyle, &c.]** into the said myne freely hee shall and

---

^{1} This phrase, "to enquire the myne," Mr. Smirke considers of Latin origin, "libertatim inquirendi mineam"—in which language he thinks the whole of the document was probably first composed.
^{2} The German miners, Mr. Smirke says, enjoyed a similar liberty. See former liberty on this head.
^{3} According to Mr. Smirke, the corresponding demand made upon the Bergmeister, by the German miners, is equally imperative, unless conflicting claims are put in, when the first finder and not the first claimant is entitled to preference.
^{4} Mr. Smirke has discovered that a like obligation was imposed on the Irons, or Iron Miners, of the forests in the ancient Earldom of Namur. He very plausibly suggests that the appellation, "Verus," by which the Dean Forest Miners designate each other, is derived from the word Firon.

shall have a dole [1] without paying anything at his first coming and shall be the last man of the Fellowship, but moreover hee shall doe coste as the Fellowship doth And if after it please the Lord to voyde he may well and if after that hit please him to come againe he may well But hee shall make Gree for the coste done in the meantyme for his p^{te} as the Fellowship can prove at the pitts mouth afterwards as another And *at* all tymes the King's Man shall come in*to* ye Myne without any **[King's man.]** Costs asking of him and shall bee the third [2] better man of the Fellowship in mayntenance and in helping of the Myne and of the fellowship But the King's Man *nor* [neither] the Lords man ought not to enter into the Myne till the pitt be gavelled (that is to say) for every dole *one* [a] penny to the King at the first [time] and after if the Fellowship doe make a new [any other] **[Pitt gavelled.]** Dole after the First Gavelling without the King's Leave wherefore for every Dole soe delivered the King shall **[King's duty]** have another Dole of the wch Mine of every Miner travelling with the said mine the king shall have every weeke a penny if soe bee that the Myner *winn* [wine] three Seames of Myne measured by the Standard *of the standard* of the King[s] of old tyme used at the least and bee it the King shall have noe more *Also* [And] the King shall have every Quarter of *a* [the] year of every Miner travelling wth in the Myne at Seame of Mine the wch is **[Gaveller's duty in receiving ye king's customes.]** called Lawe oare [3] And every weeke the Gaveller shall visitt the Tuesday the whole Mine or at [the] least within two weeks to receive the customes due to the King aforesaid And if not the Miner for the said tyme shall not bee accountable But if the Gavell^r come in the quarter to visit the Mine as is aforesaid and find not the Miner at that tyme the Gaveller shall receive soe much of [the] Mine as [it] is due to the King without leave of any Also if the Gaveller come in due tyme to receive the Customes aforesaid and the Debtor will not at that time pay then the Gaveller shall forbode soe much of myne there as

---

[1] Mr. Smirke has traced the giving of similar doles in the ancient constitutions of the Miners of Bohemia, Saxony, and the Hartz.

[2] The proportion of Profit to the Crown is found to vary in different places, sometimes being no more than a tenth part or even a twentieth or less. These provisions respecting the right of the lord of the soil, whether king or subject, have their counterparts in the old summary laws, which regulate the participation of the landowner in the discovery and working of mines; the droit de partage, or "mit-bauhalf," &c. of the German miners.

[3] See the Regard of 10 Edw. I., &c., which contains a similar specification.

hitt is due to the King by witnesse of the Miners and underneath hee shall putt a sticke of holly and after [if] the Miner carry the said Mine without gree made to the King then the Miner shall be amersed in twoe shillings and also [he] shall make Gree to the King for the Debt and if any such Mine bee forbad for Lawe Oare Then the Miner shall measure [out] soe much of the Mine that is due to the King to make Gree and the Remnant they shall carry at their own pleasure and that by the witnesse of another Miner and if hee *doth* not hee shall have the pennance aforesaid And if the Gavell^r come in due time to visitt the Mine (that is to say) Betweene Mattens and Masse ^{1} and finde not there the Miner at the end of twoe weekes (that is to say) the Tuesday in his working place as the manner is the Gavell^r shall take him that as is due And if hee bee not there present or any other for him and at what tyme the Gaveller cometh to prove if the Miner been ready to pay the Customes aforesaid or noe and they deny Then the Gavell^r by the strength of the King shall make the Miner sweare by his Faith And if the Miner bee found by his fellowship forsworne then the Miner shall be attaint **[A foresworne miner.]** against the King and shall never bee believed more agst any man and after if hee bee found with Mine within the Mine *in* [with] his cloathes pertaining to the Mine every week he shall pay to the King *one* [a] penny And the Miners of *the* beneath the wood (that **[Beneath the wood.]** is to say) Mitcheldeane Litele Deane and *River*deane [Riverdeane] every week the which the Miners travelleth in the Mine *hee* [they] shal pay *unto* the King Twelve charges of Mine by a certaine measure if they have soe much gotten by the weeke And the Gaviller shall pay the Miner there Twelve *pence* [D] Alsoe the Constable shall bee attendant by the reason of his **[Constable to keepe courts on Tuesdayes.]** office for Two weeks (that is to say) the Tuesday to hold the Court [of the Mine] that is called Myne Lawe and there to heare and [to] trye the right of our Souverigne Lord the King and of Miners and of p^{ty} and p^{ty} if any bee And at ye same Mine Lawe shall not be **[Noe foreign^r to be present.]** more sitting [but the Miners] wth the Constable but himself the Gavill^r and the Castle Clarke and the Miners before being and noe others But if soe bee [that] any other *have* [hath] to doe *with* [in] the said Mine Lawe [he shall answer for himself] and *in the said Mine Lawe noe man shall*

---

^{1} The occurrence of these pre-Reformation terms, more especially the latter, proves the original of this document to be of earlier date than that event. The portion of the day, as thus defined, would seem to answer to our forenoon.

plead neither mayntaine noe cause but onely the Miners But if soe any bee attached to answer in the said Mine Lawe **[Pleading in no other court.]** he shall answer for himself and shall be judged by the Miners of all things touching the Mine and in noe other Court and *then* hee that is found guilty Miner to Miner or any other man shall be amersed to the King in two shillings And bee it *if* [that] any will plead with any Miner for a thing touching the Miner in any other Court before a Justice or any other Man whatsoever hee bee then the Constable by the strength of the King shall require and bring the plaint into the Mine Lawe and there hit shall be tryed by the Constable and the Miners and then the party guilty shall be amersed to **[Manner of tryall by jurys by 3 degrees.]** the King as [is] aforesaid And if any plaint bee in the [said] Myne Lawe at the first day hit shall be put upon twelve Miners the wch shall give the prove the first day the Second day upon Fower and Twenty and ye third day upon eight and forty wch eight and forty shall give judgment the wch shall be affirmed firme and stable ^{1} wthout calling again for evermore And if any Miner **[Miner foresworne.]** bee found forsworne by his faith as hit is aforesaid in the proofe against any Man in the Mine Lawe Miner or Miner or Miner against any other man and the said Eight and Forty have given for judgm^t that hee is forsworne then the guilty shall be attaint against the King and shall have the pennance aforesaid and shall restore the other *of* all his *loste* [losses] in all points and never [shall] prove more Also ^{2} every Miner in his last days and *at* **[Miner may sell or bequeath his dole.]** all tymes may bequeath and give his Dole of the Mine to whom hee will as his own catele And if hee doe not *the* [his] dole shall descend to his heire and if hee to whom the dole is soe bequeathed or given by Testam^t eyther otherwise hath need to prove *his* [the] dole in ye Mine Lawe he shall come there and show the Testam^t *and* [or] bring wth him twoe witnesses to testifie the Will of the Miner and then as right is hee

---

^{1} An expression that indicates a Latin original—"judicium firmum et stabile remanebit in perpetuum absque ulla appellatione." No appeal or "calling" lies further. This appeal to successive inquests is remarkable. It resembles the process of reversing a verdict of twelve jurors by a verdict of twenty-four by the old writ of attaint. (See Blackst. Com, vol. iii.)

^{2} The German Miners Mr. Smirke found to possess these rights also. The tin-bounders of the Stannaries also bequeath their dormant liberty of mining, which is in Cornwall regarded as personal property, and passes to executors, and not to the heir.

shall bee delivered without any cost made or asked Also [1] for the customs that ye Miners done to the King the Constable that is for the time shall deliver the Miners in six weeks at the speech that is the Court for the wood before the verderers by the woodwards that keepeth the place (that is to say) Sufficient of Tymber [and] to mayntayne the King's advantages **[Timber for ye pitts and manner of haveing it.]** and profitts as also for *the* Salvaton of his Miners as they did in tyme out of mind without hurt or attachm^t made of the King's Officers (that is to say) Free the Forrest unto the Miners And also bee it that ye Miner carry tymber from the woods into his place or *into* [unto] any other the whych tymber is made and cut for the boothes for the Mine That for that *noe* [none] attachment shall be made of any man And if the Constable will deliver noe tymber as aforesaid and the Miner *of* [by] his owne authority fetch tymber in ye Forrest for the Mine and carry hit to ye Mine and after that the [said] Timber bee in *the* [their] place that is called Gavell place the wch is knowne by the old Custome Then is the tymber as their owne catele and none attachment shall be made for that Alsoe the Sea Cole Mine **[Sea cole.]** is as free in all points as the Oare Mine But if the fellowship Mine by ye weeke three charges the King shall have of every of the Fellowship a Penny Alsoe [that] noe Stranger of what degree soever hee bee but onely that beene borne and abideing within the Castle of St Brevills and the bounds of the Forrest as is aforesaid shall come wthin the Mine to see and **[A stranger not to pry.]** [to] knowe ye privities of our Souvaigne Lord the King in his said Mine Also that noe Smith holder neither Myner neither *any* [no] other shall make carriage of the said Myne *neither* by cart *nor* [neither] by waine but onely by the measure called Bill^{e}yes by ye wch the Custome of the King bee measured Soe that the Gaveller may knowe and *soe* [see] that the King have right in every *pointe* [place done] And if any such **[Measure.]** unreasonable measure shall be found then *the* [every] Miner by the strength of the King is Bayliffe to arrest the Beaste and whereof the beaste shall be forfeit to the King and ye measure burnt And bee it that the Miners for duty or for wretchedness will such wrong suffer and alsoe ye Gavell^r for his owne Lucre Then the Constable by ye reason of his office shall pursue by

---

[1] This claim to timber, at least where the forest is a royal one, has also been generally admitted into the continental mine codes. King John granted it to the tinners of Devon and Cornwall, but such a grant is now inoperative except as against the Crown.

Iron Making in the Olden Times

the strength of the King to take and to doe as is aforesaid Alsoe that noe Smith holder after he holdeth Smith or become partner to hold Smith hee shall not have none of the Franchises aforesaid within a year and a day Also by the Franchises aforesaid the Constable shall deliver *Tymber* to the Miners [Timber] sufficient to make a **[Lodges for pitts.]** lodge upon their pitt to keepe and to save *the* [a] pitt [and the mine] of the Kings and ye Miners And [1] **[Bounds of pitte.]** the pitt shall have such liberties and franchises that noe man shall come within so much space that ye Miner may stand and cast [so far from him] ridding and stones *soe farr from him* with a Bale as the manner is And shall have his marks apperteyning to his said pitt Also **[Marks.]** shall have a Bold place in the wch the Miner make and performe the tymber to build the said pitt And if any other come to travel and to work within the places aforesaid hee shall be forbode of the Fellowship of the pitt and if after hee come againe hee shall loose to the King two Shillings Alsoe ye pitt shall have a winde way [2] soe farr from him as is aforesaid pertayning to the said pitt Also the Partie that is amersed in twoe shillings shall avoid the place by the Law of the Miners Also if a pitt bee made and *upon* [be] adventure cometh another up[on] another way within the ground and drulleth [3] to the said pitt at what tyme hee drulleth to ye said pitt he shall abide till the other Fellowship of the said pitt bee present at the wch tyme if the other Fellowship will not receive him he shall *re*turne again by the forbods and by the Lawe of the Mine But if he **[Drulling a pitt.]** drulleth to the said pitt in certaine Myne then the said Mine shall bee free to both parties *which hit* [while is] dureth and afterwards [every] each one shall come agen to his owne place Saving to [every] each one ye place of others and after if one or the other doe hurt to ye other he shall restore again soe much to him if hee dig and make ye pitt fall he shall build it again and if hee

---

[1] The Mendip Miners are observed by Mr. Smirke to determine the intervening distance of their pits by a throw of "the hache" two ways, the miner standing up to the girdle in the mine groof. In Bohemia the arrow-flight fixes the limits of the work.

[2] It is presumed that "winde" in this place, and "win" or "wyne" a little further on, is the same word, viz., "win," and refers to the area or space round the pit which circumscribes the working ground of the miner, within which he is to win his ore.

[3] An original and local word. It seems to be allied to drill a hole. (I do not think the word strictly local. Thrull, drill, thrill, thirl, and thurl, are all current elsewhere—all from Saxon διηlιαη.)

distrouble the other *soe* that he may not travaile to *win* [wyne] his proffitt and the Customes of the King hee shall restore all the lost of the king and the Miner Alsoe if any bee wrongfully forbode by the Miner or by **[Wrong forbode.]** any other Then hee that is forbode shall come to ye pitt and shall bring wth him his Instruments pertaining to ye Mine with his light as another of ye Fellowshipp and the*re* [then] hee shall abide so long as the fellowshipp and then by *the* judgment of eight and forty he shall receive so much as any other of the Fellowship &c.

[The miners' names.] John Garron, Stephen Preest, John Clarke, Thomas Wytt, Thomas Norton, John Hathway, Thomas Michill, John Mitchill, John Smith, John Lambert, Nicholas Orle, John Barton, Richard Haynes, John Armiger, Walter Rogers, Richard Hathen, Walter Smith, William Miller, Thomas Cromhall, Walter Dau, [John Loofe, Roger Shin, Henry Norton, Thomas Forthey, Walter Waker,] Richard Timber, William Baker, Thomas With, John Baker, Phillip Dolewyer, John Adys, William Hynd, William Tallow, John Brute, John Mitchill, Richard Hopkins, Thomas Baster, John Laurence, Thomas Tyler, Walter Dolett, William Callowe, Richard Holt, Walter Warr, John Robert, Henry Doler, John Parsons, William Holder, Thomas Clarke. [1]

Be it knowne to all men that shall see *or* [and] heare this writing that the Inquest of fforty and eight Miners witnesses and confirmeth all the Laws comprized in ye said Roll for witnesse whereof they have put their Seales.

Written out of a parchm^t roll now in
ye hands of Richard Morse of Clownwall
7 Jany 1673

THO. DAVIES.

Memor^{dm} this was afterwards printed for W^m
Cook at the Pelican in Little Britain hoc Titulo
    The Laws & Customes of the Miners
    in ye Forest of Dean
    The Rules & Orders of St Brevaills
        Court endorsed
        Written 7 Jany 1673.

---

[1] Of course there should be forty-eight signatures, as appended, doubtless, to the original document. Probably some of them had become illegible, and therefore were omitted altogether by the copyist of 1673.

Principles of Mining - (With index and illustrations)Valuation, Organization and Administration. Copper, Gold, Lead, Silver, Tin and Zinc.
Herbert Hoover
Oxford City Press, 2011
216 pages
ISBN: 978-1-84902-408-2 Hardback
ISBN: 978-1-84902-438-9 Paperback

Available from www.amazon.com, www.amazon.co.uk

Herbert Hoover was a Mining Engineer and author before becoming President of the United States of America during the great depression of the 1930s. This book is fully illustrated and comes with an index. In the Preface, Hoover describes the book in this way: "This volume is a condensation of a series of lectures delivered in part at Stanford and in part at Columbia Universities. It is intended neither for those wholly ignorant of mining, nor for those long experienced in the profession."

The Cliff Ruins of Canyon de Chelly, Arizona - with original illustrations and index - Sixteenth Annual Report of the Bureau of Ethnology to the Secretary of the Smithsonian Institution, 1894-95, Government Printing Office, Washington, 1897, pages 73-198
Cosmos Mindeleff
Benediction Classics, 2011
180 pages
ISBN: 978-1-84902-387-0

Available from www.amazon.com, www.amazon.co.uk

The Canyon de Chelly is one of the best Cliff Ruins regions in the United States. This book details the pueblo dwellings in the region, with over a hundred black and white diagrams and photographs. The original index and footnotes have been preserved.

Shelters, Shacks and Shanties - with 1914 cover and over 300 original illustrations
D.C. Beard
Benediction Classics, 2011
200 pages
ISBN: 978-1-84902-320-7

Available from www.amazon.com, www.amazon.co.uk

"Shelters, Shacks, and Shanties" was written and illustrated by Daniel C. Beard in 1914. He was one of the founders of the Boy Scouts movement in America. The book explains how to build shelters, from the simplest requiring a hatchet, to elaborate constructions such as a homestead. The book is addressed to "boys of all ages". This version comes illustrated with Beard's 338 pen and ink drawings, as well as a cover based upon the original 1914 cover design. The book tells you how to build: Sod Houses Log Cabins Over-Water camps Railroad Tie Shacks Navaho Hogans And it also explains: How to Use an Axe How to Split Shakes How to Build a Fireplace How to Make Doors & Latches

Manual of Military Training - Second, Revised Edition - with Index, footnotes and copious images
James A. Moss
Benediction Classics, 2011
896 pages
ISBN: 978-1-84902-303-0

Available from www.amazon.com, www.amazon.co.uk

The Manual of Military Training was originally published in 1914, and intended as a complete training guide for a company. The book was used for cadets in military schools, the Organized Militia and also for company officers of the Regular Army. This version comes fully illustrated and with a copious index. It is a must buy for boys of all ages.
Also from Benediction Books …
Wandering Between Two Worlds: Essays on Faith and Art

Anita Mathias
Benediction Books, 2007
152 pages
ISBN: 0955373700

Available from www.amazon.com, www.amazon.co.uk

In these wide-ranging lyrical essays, Anita Mathias writes, in lush, lovely prose, of her naughty Catholic childhood in Jamshedpur, India; her large, eccentric family in Mangalore, a sea-coast town converted by the Portuguese in the sixteenth century; her rebellion and atheism as a teenager in her Himalayan boarding school, run by German missionary nuns, St. Mary's Convent, Nainital; and her abrupt religious conversion after which she entered Mother Teresa's convent in Calcutta as a novice. Later rich, elegant essays explore the dualities of her life as a writer, mother, and Christian in the United States-- Domesticity and Art, Writing and Prayer, and the experience of being "an alien and stranger" as an immigrant in America, sensing the need for roots.

About the Author

Anita Mathias is the author of *Wandering Between Two Worlds: Essays on Faith and Art*. She has a B.A. and M.A. in English from Somerville College, Oxford University, and an M.A. in Creative Writing from the Ohio State University, USA. Anita won a National Endowment of the Arts fellowship in Creative Nonfiction in 1997. She lives in Oxford, England with her husband, Roy, and her daughters, Zoe and Irene.

Anita's website:
    http://www.anitamathias.com, and
Anita's blog Dreaming Beneath the Spires:
    http://dreamingbeneaththespires.blogspot.com

The Church That Had Too Much
Anita Mathias
Benediction Books, 2010
52 pages
ISBN: 9781849026567

Available from www.amazon.com, www.amazon.co.uk

The Church That Had Too Much was very well-intentioned. She wanted to love God, she wanted to love people, but she was both hampered by her muchness and the abundance of her possessions, and beset by ambition, power struggles and snobbery. Read about the surprising way The Church That Had Too Much began to resolve her problems in this deceptively simple and enchanting fable.

About the Author

Anita Mathias is the author of *Wandering Between Two Worlds: Essays on Faith and Art*. She has a B.A. and M.A. in English from Somerville College, Oxford University, and an M.A. in Creative Writing from the Ohio State University, USA. Anita won a National Endowment of the Arts fellowship in Creative Nonfiction in 1997. She lives in Oxford, England with her husband, Roy, and her daughters, Zoe and Irene.

Anita's website:
    http://www.anitamathias.com, and
Anita's blog Dreaming Beneath the Spires:
    http://dreamingbeneaththespires.blogspot.com

 www.ingramcontent.com/pod-product-compliance
Ingram Content Group UK Ltd.
Pitfield, Milton Keynes, MK11 3LW, UK
UKHW040205230326
11407UKWH00001B/5